Geo Power

Stay Warm, Keep Cool and Save Money
with Geothermal Heating & Cooling

Donal Blaise Lloyd

Foreword by Lawrence A. Muhammad

PixyJack Press INC

Published by PixyJack Press, Inc.
PO Box 149, Masonville, CO 80541 USA

print ISBN 978-1-936555-58-1
Kindle ISBN 978-1-936555-59-8
epub ISBN 978-1-936555-60-4

Library of Congress Cataloging-in-Publication Data

Lloyd, Donal Blaise.
  Geo power : stay warm, keep cool and save money with geothermal heating &
cooling / Donal Blaise Lloyd ; foreword by Lawrence A. Muhammad. -- First edition.
     pages cm
  Includes index.
  Summary: "Covers geothermal (ground-source) heat pumps for residential
heating and cooling, including system options, costs, payback issues, performance
standards, and installation concerns. Also explains the science and technology of
heat pumps and how they work"-- Provided by publisher.
  ISBN 978-1-936555-58-1
  1. Ground source heat pump systems. I. Title.
  TH7417.5.L637 2015
  697--dc23
                  2014042301

Earlier Edition: *The Smart Guide to Geothermal,* © 2011

Back cover photo courtesy of Lawrence A. Muhammad
Geothermal illustrations by Will Suckow.
Book design by LaVonne Ewing.

More info at: www.GeoPowerful.com

To Deborah and Nancy

Contents

Part II — Unraveling the Science & Technology of Ground-Source Heat Pumps

Part III — The Broad View of Geo Power and Its Opportunities

Appendix

List of Illustrations

Foreword

How do we tap into geothermal power, a plentiful energy source? This important book provides the answers. It gives clear explanations of the concepts, the laws of thermodynamics, and the applications that are germane to ground-coupled systems which utilize Earth's natural and free heat energy to heat and cool our homes and businesses.

For the last couple of centuries we have been using precious fossil fuels stored in the earth for HVAC/R (heating, ventilation, and air conditioning/refrigeration). In the past couple of decades, however, geothermal—"heat from the earth"—for HVAC/R has made its mark as a viable alternative: it has become a primary energy source for controlling temperatures in conditioned spaces, large and small, thus vastly reducing the operating costs of new or retrofitted homes, schools, health care facilities, and office buildings throughout the United States, Canada, Europe and the world.

As a national trainer for the International Ground Source Heat Pump Association (IGSHPA), a professor at several Community Colleges, and a licensed builder in Michigan, I have come to know the ins and outs of the geothermal installation business. The book you have in your hand is used in my entry-level courses and workshops because it is easy to understand and answers the myriad of questions that geothermal HVAC/R novices may have, as well as those who wish to pursue lucrative related businesses.

In the early years of developing a "geothermal curriculum"

I was forced to use textbooks that did not have a whole lot to say about geothermal for HVAC or were too esoteric or scientific for the beginning student. However, I continued to search and found author Don Lloyd's book. I read it within 24 hours and immediately got it approved for coursework at Wayne County Community College District in Michigan. His books have been used ever since in the schools and are also distributed to apprentices at projects throughout the Midwest and the East Coast.

Geo Power (the expanded edition of *The Smart Guide to Geothermal*) is written so that, in a short period of time, my students and trainees are able to have the knowledge and understanding of not only why geothermal can be used, but the scientific reasons and evidence for its effectiveness as well.

Whether you are an existing homeowner looking for an environmentally safe and sustainable alternative to fossil fuels, or a business professional responsible for improving your company's bottom-line operating costs, this is the right book at the right time to read and enjoy as you learn more about geothermal principles, applications, and the "science" behind it all...as we each must do our share in making sure that future generations have safe and clean energy sources delivered to their homes while not compromising our comfort—or budget—today.

Professor Lawrence A. Muhammad

Preface

The next generation of home heating/cooling is here! I can even foresee traditional oil burners and gas boilers becoming nearly obsolete in the future. I'm pleased to say there is an excellent alternative to fossil-fuel burning, home-heating systems and it is the geothermal heat pump (GHP).

What is the driving force behind the growing GHP industry? It is the ability to save money, year after year. And how does the homeowner or building owner take advantage of this? The answer: successful installations by qualified installers.

To be successful, the first step for installers—and anyone in this trade—is not just understanding what homeowners care about and how to best address their needs, but also having a good grasp of the science and technology fundamentals.

How does a ground-source heat pump residential system start with perhaps a 50°F ground temperature and increase the temperature and the heat energy levels to produce a comfortable zone throughout the home without burning fuels? All it takes is the application of some simple laws of nature.

I addressed this to a limited extent in my first book for homeowners: *The Smart Guide to Geothermal: How to Harvest Earth's Free Energy for Heating and Cooling.* My goal then was to provide all of the information homeowners need to make that important decision about installing a geothermal heat pump in their homes, including the various types of systems, financial investment and payback times, and finding a contractor.

John Geyer, a West Coast geothermal expert/consultant put it this way to me: "The appeal and value of *The Smart Guide to Geothermal* is its high-content density and dearth of tech-speak. It teaches users enough to trust geothermal and to want its benefits. When a homeowner picks up enough talking points to chat it up, he/she becomes the industry's primary sales force."

When giving presentations on geothermal, however, I was asked for a more complete and systematic explanation of how a GHP really works. My discussions with Lawrence Muhammad, a certified trainer and community college professor, further emphasized this need. It became clear some my explanations needed more depth.

So I searched for an overall description that covers this science in detail; surely something must be available since the same technology is used in refrigerators and air-conditioners. Well, I found only bits and pieces that had to be put together in the context of a heat pump, or even more specifically, into demystifying the refrigerant loop of the heat pump. Thankfully it was not quantum mechanics, but a set of natural, physical laws observed in everyday life.

My goal: make it simple, easily understood science... for myself and for those who sell and install these systems. And so, this book was born with an expanded Science & Technology section.

Of course, not everyone needs or wants this level of information, but many heat pump installers feel they should be aware of what is going on behind the scenes. It is like a doctor knowing his patient. So for those teaching heat pump systems in community colleges and other venues, as well as heat pump trainers, designers, and future installers, here is an easy-to-grasp, concise look at how GHP systems achieve such impressive efficiency ratings.

And if you are a curious homeowner, builder or architect, by all means, read on!

This book is organized into three parts:

Part I—The Big Picture is a general description of geothermal heat pump systems (including my personal experience), the advantages, disadvantages, costs, payback time and increased market value of the home. This section also covers finding a contractor, living with a GHP system, and gives examples of homes with installed geothermal systems.

Part II—Unraveling the Science and Technology section goes into more technical depth in describing the flow of heat from the ground to the house, the "magic" of the compressor, how 400–500% efficiencies are obtained, and how the Laws of Thermodynamics come into play.

Part III—The Broad View looks at the global marketplace, commercial / institutional geothermal systems, future trends, and career opportunities.

The appendix includes references and links to other geothermal websites and state energy offices, plus a listing of companies that manufacture ground-based residential geothermal heat pumps in the United States and Canada.

We start out locally (a few feet below the ground in our backyard) and end with a global outlook—all with the same goals of lowering heating costs, reducing pollution with renewable sources, and creating energy independence.

For those of you on the front lines of this proven technology, it is an exciting time. May this book prove to be a helpful tool in your toolbox.

Don Lloyd

It Is More Efficient to Transfer Than Burn

I was reading a book recently about a family having a dinner conversation. One of the young daughters was discussing a history-class subject where in past times people had killed whales in order to provide oil for burning in lamps. "What a terrible waste!" she said. I think that our yet unborn great-grandchildren will look back on our times and also say, "They actually *burned* oil? What a terrible waste!" All generations do what they must in order to procure the energy they need, whether it is whale oil, coal, gas or petroleum, but there is now a smarter way, an alternative.

The next generation of heating systems using proven refrigerator technology is readily available for homeowners—it is called ground-source, geo-exchange, earth-coupled, or geothermal heat pump (GHP). This technology has been slowly perfected over

According to the Environmental Protection Agency, geothermal ground-source heat pump systems are one of the most energy efficient, environmentally clean, and cost-effective space conditioning systems available.

the last 50 years to the point where GHPs are now becoming
mainstream technology for domestic heating and cooling, and
indeed for commercial and institutional applications.

It is far more efficient, less costly and makes more sense to
transfer existing heat energy from a source that is renewable:
the earth. It is not necessary to go deep to capture heat from the
molten center of the earth. GHPs capture heat energy just below
the frost line in the backyard; heat energy which is continually
renewed by the sun as well as the earth's center. As long as the
sun shines (still 5 billion years to go!), it will pump heat energy
year-round into the ground beneath our feet. And we can tap
into that easily, right now.

My research was conducted to make certain that this book
contains not only everything you need to know about ground-
based geothermal heat pumps, but also a lot more that you may
be curious about. Embedded in the sum of all these chapters is
one single concept that is a *must* for you to understand. It is the
central reason for a geothermal heat pump to even exist. It is the
reason they are being installed in increasing numbers worldwide.
It is the reason subsidies and tax credits are being provided in
the United States and Canada. It is the reason I installed such a
system in my own home...

**A GHP is superior to any other method of home heating
and cooling in three very important ways:**

1) Does not combust any fossil fuels
2) Uses a renewable resource
3) Has the lowest cost of operation

Figure A-1 on the next page puts that all together.

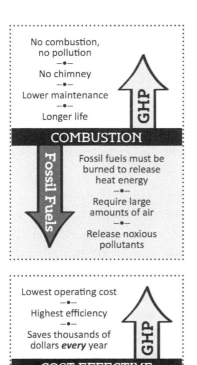

No combustion,
no pollution
—•—
No chimney
—•—
Lower maintenance
—•—
Longer life

COMBUSTION

Fossil fuels must be
burned to release
heat energy
—•—
Require large
amounts of air
—•—
Release noxious
pollutants

Lowest operating cost
—•—
Highest efficiency
—•—
Saves thousands of
dollars *every* year

COST EFFECTIVE

All cost more to
operate than a
geothermal heat pump
—•—
All require annual
maintenance, which
costs more money

**Geothermal
Heat Pump Systems
Compared to
Fossil Fuel-Based Systems**

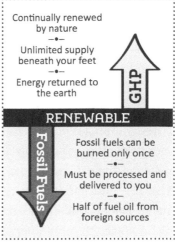

Continually renewed
by nature
—•—
Unlimited supply
beneath your feet
—•—
Energy returned to
the earth

RENEWABLE

Fossil fuels can be
burned only once
—•—
Must be processed and
delivered to you
—•—
Half of fuel oil from
foreign sources

Figure A-1
In all cases, a geothermal heating/
cooling system comes out on top. It's
almost like having your own utility.

GHP = *Geothermal Heat Pump*
Fossil Fuels = *fuel oil, natural gas,
propane, coal*

1) No Combustion of Fossil Fuels

Let's face it; we have been heating our homes with almost Stone Age technology. Burning fossil fuels (using combustion) means flames, and much of the heat energy from flames is wasted. At the same time, combustion produces pollutants, including deadly carbon monoxide.

Fuel oil, coal, natural gas, propane, cord wood, and pellets must all be burned to release their heat energy. Burning also takes large amounts of oxygen out of your home and requires a chimney for venting the exhaust (except a few non-vent propane heaters). Electric baseboard heating does not require burning within the home, but we must remember a significant amount of electricity used in the United States is generated by coal burning plants—plants that are only 35–45% efficient, many of which are still equipped with towering, belching chimneys.

In contrast, a GHP system produces not a single spark. It offers emission-free operation on site. Heat energy is transferred; nothing is burned. No chimney is required. Heat pumps are among the few reliable and widely available heating technologies that can deliver thermal comfort with zero emissions. The IEA (International Energy Agency) Heat Pump Centre has stated that heat pumps are one of the most significant available technologies that can offer large carbon dioxide (CO_2) reductions. In addition, GHPs require no on-site fuel storage. With less system stress, a GHP has a much longer life and requires less maintenance. (*Chapter 9 – Learning to Live With a GHP System* also provides actual experience in this area.)

That is one star for a GHP: ✪

2) Uses a Renewable Resource

You can only burn fossils fuel once, but a GHP system transfers heat from the earth; heat that is forever renewed by the sun and the earth's center. As for fuel oil, coal, natural gas (NG), and propane (LP), none of these fuels are renewable. All are fossil fuels and all release pollutants into the air when burned. Figure A-2 shows the percent of housing units by heat type from US Census data. No GHP in the census list. Maybe someday we can change that.

Type of Heating in Occupied Housing Units

Figure A-2
"Other" refers to coal, wood, kerosene, solar and none.

Source: US Energy Information Administration

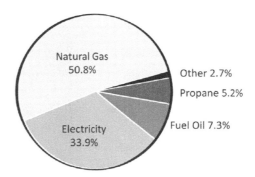

Natural Gas 50.8%
Other 2.7%
Propane 5.2%
Fuel Oil 7.3%
Electricity 33.9%

Propane, also known as LP gas or LPG, is derived from petroleum. It is not as clean as natural gas, but it can be delivered by truck as a liquid under pressure; no extensive underground pipe infrastructure is required. The US Census Bureau states that in 2013 LPG heated 8.0 million US homes.

Natural gas (NG) is 70% methane, which is 25 times worse than CO_2. It is shipped in tankers and stored as LNG, liquefied natural gas. NG is the cleanest of the fossil fuels, but still releases CO_2, nitrogen oxides, and sulfur oxides. In 2013 approximately 57.2 million (or 51%) US homes use NG. Hydrofracking, although quite controversial, has increased supply to the point where the US is preparing to liquefy and export it.

Fuel oil: In 2005, the US imported 65% of our oil. Today it is less than 50% due mainly to hydrofracking plus lower demand. The Census Bureau reported in 2013 that 7.3 million US homes are heated with fuel oil, mainly in the northeast.

Coal, a readily combustible sedimentary rock consisting primarily of carbon, emits carbon dioxide emissions during burning that are double that of NG. It is by far the most abundant fuel produced in the United States. Anthracite coal, the hardest type, is used for home heating and it is cheap. Coal burning, self-stoking stoves are available, but they do create ash and

---

Homes with Oil Burning Furnaces

I must confess that the low number of oil users came as a great surprise to me. In part because of corporate moves, my wife and I have owned homes in Maryland, Minnesota, Massachusetts and New York State. Every one of those homes had oil burners. I had a mind-set that most homes, except those of city dwellers, burned oil. But that is not true.

Home heating sources have changed over the past 50 years. In 1960, 32.5% of US households heated with fuel oil, but only 7.3% in 2009. On the other hand, electricity was used for heat in only 1.8% of homes in 1960, but 33.9% used electricity in 2009. Propane usage stayed about the same, but natural gas heating has grown from 43.1% to 50.8%. Coal dropped from 12.2% to 0.1%, thankfully. SOURCE: *U.S. Energy Information Administration*

---

dust. Coal is a messy fuel, but can be an option, especially if you live near the eastern Pennsylvania anthracite area. However, many of those coal-burning customers are finding that in spite of having 7 billion tons of anthracite coal underground, mines are closing down, causing shortages.

As far as the number of homes heated by coal, the US Energy Information Administration (EIA) says that 0.1% or only 100,000 homes were heated with coal in 2009, down from nearly 6.5 million in 1960.

Wood (pellets and cordwood): All are renewable, of course. Most homes use them as auxiliary or backup heat. In 1960, 4.2% of US households heated with wood, but that number has dropped to 1.3% in 2007 (US EIA).

Electric baseboard: Sources for our electricity include natural gas, coal, oil, nuclear and renewable energy *(see Figure A-3)*, but even these sources vary greatly from region to region, utility to utility. Today approximately 36.1 million homes are heated by electricity, one third of the total. It is interesting to note that 60% of these are in the South.

In summary: It requires a small amount of electrical energy to transfer a large amount of heat energy in a geothermal system. How this takes place is described in *Chapter 3 – How a GHP Works.* So far, that is another star, adding up to two for a GHP: ✪✪

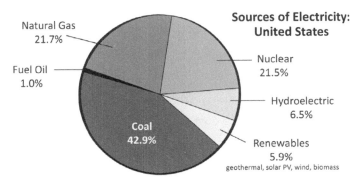

Source: US Energy Information Administration, Annual Energy Review 2013

Figure A-3

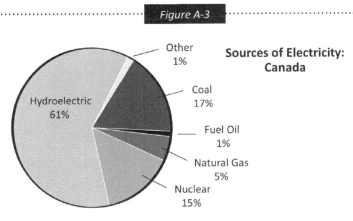

Source: Canadian Office of Energy Efficiency, 2008

3) Lowest Cost of Operation

You have heard the expression, "Follow the Money." Well, this is where the money is. The cost of operation for a heating system can mean a difference of thousands of dollars per year, *every* year. And the factor that most heavily influences that cost is the efficiency of converting the fuel into heat energy.

Every fuel source has a different basis or unit—gallon, ton, kilowatt-hour (kWh), etc. The only way to compare the cost of operation is to define the heat content in each of those units in terms of a common energy unit, such as Btu (British thermal unit). This provides a dollar value per Btu (actually, it is handier to use dollars per million Btu). Finally, we must modify that cost-per-Million-Btu (MMBtu) by the actual efficiency of converting

Heating Cost Comparison Data

Fuel Type	Fuel Unit	Fuel Price per Unit	Btu per Unit	Fuel Price per Million Btu	Efficiency	Fuel Cost per Million Btu
GHP	kWh	$0.116	3,412	$33.90	400%	$8.48
Coal	Ton	$200.00	25,000,000	$8.00	75%	$10.67
Pellets	Ton	$250.00	16,500,000	$15.15	68%	$22.28
Wood	Cord	$200.00	22,000,000	$9.09	55%	$16.53
Propane	Gallon	$1.93	91,333	$21.16	78%	$27.13
Natural Gas	Therm*	$1.25	100,000	$12.50	78%	$16.03
Electric Baseboard	kWh	$0.116	3,412	$33.90	100%	$33.90
#2 Fuel Oil	Gallon	$2.33	138,690	$16.80	78%	$21.54

*Therm = 100,000 Btu

Figure A-4
A GHP (geothermal heat pump) has the lowest cost per BTU of any other heat source. Source: US Dept of Energy, 2008

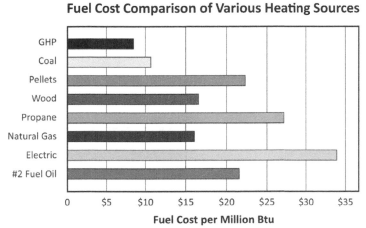

Fuel Cost Comparison of Various Heating Sources

Figure A-5
A GHP has an even lower cost per BTU than coal. Source: US Dept of Energy, 2008

a particular fuel source into heat energy. Then we have a *real* cost per MMBtu.

Figure A-4 shows how the US Department of Energy chart computed the fuel cost per MMBtu for each type of fuel. Their conclusion is very clear: **A GHP has the lowest cost per energy content than any other fuel type.** This point is even more obvious when the data is converted to a bar graph as shown in Figure A-5.

This cost-savings is so important it is worth two more stars for a GHP with a grand total of four: ✪✪✪✪

Using EIA data, we can attempt to make predictions about the future:

1) Oil prices went down in 2014, reducing the cost per million Btu of heating oil, but EIA predicts this will slowly increase.
2) Natural gas prices will go up as lower fuel oil price make hydrofracking more risky.

3) Electric rates will continue to increase less than 5% per
   year, but **geothermal heat pump** efficiencies will increase
   even more as GHP components continue to steadily improve
   and as the US EPA continues to mandate increased future
   efficiency standards.

The results of this projection show that GHPs will have an
increasing advantage over time. Two factors are involved here.
One is the cost of fuel, which is assumed to be increasing for all
types. The other factor is the efficiency of converting the fuel
to heat energy. In the past few years, high-efficiency oil burners and gas boilers have become available, but GHP efficiencies
have not yet peaked and are still increasing. In time, newer GHP
components will be able to do the same work with less electricity. That is why you see in Figure A-6 a GHP cost advantage will
be even better in the future.

The trend is interesting. It shows the GHP relative cost over

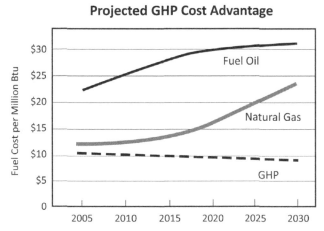

Figure A-6
Future GHP Systems will increase their cost advantage.

time compared to natural gas and heating oil. The result, based on the above assumptions, shows an increased future advantage.

So, if a GHP is so talented and the most modern system, why is it not *the* choice for every residence on the continent? The answer in part is due to lack of knowledge, as covered in more detail below. But to be fair, there are disadvantages to GHP systems. The most obvious is that the total cost to install a GHP will usually be higher than a conventional heating system because of the need for ground excavation or drilling. Even that disadvantage is not necessarily bad news. *Chapter 13 – Payback: The Real Bottom Line* includes some interesting revelations.

The Need for a GHP Book

All of this information is useless if no one is aware of it. Ground-based geothermal heat pumps are underutilized because they are not well known or are misunderstood. My architect felt that a GHP was beyond my budget. My builder had never built a home with a GHP. But the biggest problem was my bank. The bank construction-loan underwriter did not fully understand this technology. All were obstacles that had to be overcome by education and persistence.

I am not alone in this. A 2008 Oak Ridge National Laboratory study stated: "Lack of consumer knowledge and/or confidence in GHP system benefits is one of the key barriers to growth of the GHP industry." The Canadian GeoExchange Coalition (CGC) acknowledges a lack of public knowledge by including in its charter a goal of "increasing awareness."

In addition, a recent survey reported that 9 out of 10 people know what a solar panel is, while only 2 out of 10 know what a geothermal heat pump is.

From 2005 to 2008, shipments of GHPs in the United States doubled (121,243 units in 2008), in great part due to the surge in oil prices. But in 2008 alone, new housing starts in the US plus Canada totaled over 1.1 million. As you can see, GHP shipments are still a very small part of the total.

While this book has a North American perspective, the material is international in application. In fact, we have been falling behind in the application of this technology compared to Europe and Asia.

We must find a way to capture public and political imagination. This book is a modest attempt to change that, and only one voice singing the praises of a different approach. It would indeed be cool and to the benefit to all countries if that voice swelled into a quartet, then a chorus, and even into a grand orchestra. ☼

....

Different Terminology; Same Technology
Geothermal Heat Pump (GHP)
Ground-Source Heat Pump (GSHP)
Geoexchange (GX)
Earth-Coupled
Earth Energy
....

Part I

The Big Picture:

What Geothermal Heat Pumps Can Do for Homeowners

Chapter 1

A Different Approach to Heating and Cooling

What are you paying to heat your home this year? Responses to that question generally run from $2,000 to $5,000 per home. The Institute of Social and Economic Research (ISER) conducted a study in Alaska where the average 2,500 square-foot (232 sq m) home paid $4,500 in 2008. According to the EIA, most of the United States will experience higher home heating costs due lower than average winter temperatures. In 2014-15, oil costs will drop 6.4%, but natural gas prices will increase nearly 6% and electricity will go up by about 3%.

My last home, a renovated two-story brick, 8,000 square-foot (743 sq m) former schoolhouse, had an annual bill in 2005 of over $7,000 that has since grown much higher. This was the major reason for selling it, even though my wife and I had put our hearts and souls and many years of our lives into that building.

At that time, I had a very real concern about how much I would pay for fuel in the future. But now I have no uncertainty about *my* fuel costs: zero! Why?

Fuel oil and propane were two options where I could not control prices. Natural gas was not even available. But I could do without burning fossil fuels and so I took a different approach. I installed a ground-based geothermal heat pump (GHP) in our new home. Designing and building our home was a four-way collaboration between our architect, our builder, my wife and myself. But the geothermal decision was my passion; my responsibility for better or worse.

What I envisioned was a heating system that would provide comfortable heating and full-house air-conditioning (AC), did not have to burn anything, never needed the nozzles cleaned (it had no nozzles), emitted no noxious gases up the chimney (no chimney at all), provided partial free hot water, had lower operating and maintenance costs than an oil burner, was far more efficient and quiet, and finally, saved thousands of dollars every year. And that is exactly what I got! Although it is not a common situation, the installation of my particular system cost less than an oil burner plus full AC.

The US EPA states, "Geothermal heat pump systems are the most energy efficient, environmentally clean and most cost effective space conditioning systems available." But more importantly, this underutilized technology has the potential to be an important factor in reducing every country's dependence on foreign oil while at the same time reducing air pollution.

What is a geothermal heat pump (GHP)? In short, a GHP draws heat from the ground. An air-source heat pump, on the other hand, uses air as the source of heat. Air-source systems can be useful in warmer climates, but are limited in efficiency and application. When comparing air-source and ground-source heat pumps, GHPs are quieter, last longer, need little maintenance, and do not depend on the temperature of the outside air.

Other geothermal heating systems use naturally heated underground water, such as hot springs or reservoirs. These direct-use geothermal systems are not included in this book.

You should also note that a GHP system is not the same as geothermal energy or geothermal electric power (GEP). GEP is an electrical power generating system that includes drilling deep into the earth's crust to tap into heat that can be used to turn a turbine to generate electricity *(see below)*.

A ground-based GHP uses heat just below the frost level of the ground. This heat energy is constantly renewed by the sun and by heat from the earth to provide a relatively constant temperature. The GHP takes heat from the ground in the winter to heat a home, and in the summer, it provides cooling by putting heat back into the ground. So, we are merely borrowing heat from the earth.

This technology turned out to be a very personal thing with me. Installing a geothermal heating system in my new home instead of an oil burner was not as easy as I had imagined, but clearly well worth it! ✪

Geothermal Electric Power

There is another, but quite different geothermal effort underway that is going to have a large future impact on our energy supply. You should be aware of this because in some ways it is often confused with geothermal heat pumps. Geothermal electric power (GEP) is similar to the geothermal heat pump in that it takes heat energy from the earth, but there are major differences. First, it is not for residential applications; it will be closer in size to a private utility operation. Second, while a GHP transfers heat energy from the earth to warm a home (or building), the GEP converts heat energy into electrical energy. The geothermal power system captures heat energy that originates at the earth's molten center—"using the earth's furnace."

The crust of the earth is made up of huge plates that are in constant but very slow motion relative to one another. Geologic processes allow molten magma to rise up near to the surface, studio. There was wall space for hanging very large canvases. We had fun living there, with many Boston and New York City visitors, among them art collectors, including two Boston luxury hotels, who bought my wife's work. Other artists from the area were invited to use the upstairs classrooms as art studios.

We were able to put the building on the National Register of Historic Buildings and I finally published the full story in a memoir, *Snake Mountain Trilogy: A Berkshire Memoir.* That led to a book party in the schoolhouse's former second-grade room (by then our dining room) where I told our guests, "We had our building closing on Halloween Eve, we moved in on Halloween Eve, and today is Halloween Eve. I want you to know that you are the largest, and by far the oldest trick-or-treaters we have ever had." I also explained that this was the first time I had ever publicly used my full name—Donal Blaise Lloyd. As a fourth grader, I was teased, "Blaise, Blaise, your pants are on fire!" Those kids were not only annoying; they could not even spell. And I had a chronic problem with my first name because so many insisted

After years of labor and love, our schoolhouse became our home, but heating costs were a real shocker.

on adding a "d" at the end. And when corrected, they made it worse by putting the accent on the second syllable. But I decided it was time to take charge of my name and do it my way.

But then came the bad news, really bad. The building required 3,000 gallons (11.3 kl) of fuel oil every season. At 99 cents a gallon it was bearable, but price increases struck soon and hard! When the cost passed the $7,000 level with every indication of going higher, it was clear that we had to sell it, as difficult as that was. So we did.

The new owner, an artist who fully appreciated the artist's stewardship of the building, hired an engineering consultant to determine if a geothermal heat pump was practical considering the age, construction and size of the building. This was a new and intriguing concept. The more I learned about geothermal energy, the more hooked I became.

A Geothermal Heating System for our New Home

Starting all over again, I was determined that I was not going to be an oil captive any longer. Our new, smaller home was going to have a ground-based geothermal heating/cooling system. I did my homework to understand how GHP systems worked, their advantages and disadvantages, the types of systems available, and then studied the companies that manufacture the heat pumps. I learned about efficiencies and costs, and had to relearn the physics that makes the whole thing work.

Have you ever had the feeling that something you're considering is absolutely right? It may be a small point or a large, important item, and it does not happen often, but once in a while something clicks. It just feels right. And so it was with this endeavor for alternative heating.

We found exactly the right architect to translate our needs

3) If fire is a concern, consider that with a GHP we would not be burning anything, therefore fire or carbon monoxide gases will be highly improbable.

4) It would be a sad thing if this bank went against our national need to reduce our dependence on foreign oil. I quoted the EPA: "If one million homes used geothermal heat pumps instead of oil, 21.5 million barrels of crude oil would not have to be imported each year."

They backed down. Construction started, was completed and we moved in. Was my pushiness worth it? Absolutely! We had no fuel bills. It was one of the smartest decisions I ever made. And the higher oil prices go, the smarter I seem to get.

All of my initial problems resulted from lack of knowledge. Many people I talk to have never heard of geothermal heating, or have incorrect information. It is an underutilized technology. Well over a million housing units are started each year in the United States, but only 120,000-plus geothermal heat pumps were shipped in 2008, according to the Department of Energy, and that includes not just homes, but schools and commercial buildings. How many of those new homeowners are even aware of what can be done? ✪

....

If one million homes used geothermal heat pumps instead of oil, 21.5 million barrels of crude oil would not have to be imported each year. Source: EPA

....

Germantown, New York

Earth Smart Home

Henry Hudson recorded his landing at what is now the Ro-Jan Creek in Germantown on his way up the Hudson River in 1609 to get fresh water. In 1709, the Queen of England convinced hundreds of Palantine Germans to settle here in order to produce pitch for wooden ships.

Germantown today is a small, interesting, rural town on the Hudson River of about 4,000 residents and 32 square miles (83.5 sq km). The Catskill Mountains, on the other side of the river, are very visible to the west.

This is where we found a partially built shell of a home on a former apple-tree farm. The building had just barely been started, but due to a divorce, was never finished. It was just left as an eyesore. When my wife, Martha, received her MFA degree at Bard College a few years before, she never dreamed that we would someday be living within a few miles of that campus.

Details: Single-story home, essentially new construction, 2,400 square feet (222 sq m), two-zone heating/cooling, 4-ton ECONAR GHP open-loop system with the main water well as heat source, and ductwork for hot and cold air delivery. The work was completed in 2007.

Open loop means that instead of having a water loop in a continuous cycle picking up heat and dropping it off, water from a source such as a well is piped into the GHP, releases its heat and is then piped back into the well again. The advantage is a much lower installation cost since the excavation for the pipes must be done for the water supply anyway. The disadvantage is a higher operating cost since the submersible well pump runs more often.

We found an architect in nearby Massachusetts and relayed our needs and wants. Needs: an attached art studio of 800–1,000 square feet (74–93 sq m), a ground-based geothermal heating/cooling system, and a single-floor, open interior with exterior views. Wants: a garage with a steel roof. When the final costs came in, some reductions had to be made: the garage was dismissed and other changes made, but the GHP and the studio were firmly in the plan. The bank resisted the GHP, but wilted when challenged.

Do we love it? Definitely yes! We have no fuel-oil bills. We have very comfortable heating and cooling plus partially free domestic hot water.

Chapter 3

How a Geothermal Heat Pump Works: Borrowing Heat from the Earth

I f you dig down about five feet or so in the ground to below the frost level, you will find the ground temperature to be amazingly constant, 40° to 70°F (4°–21°C), depending on the location.

It is cooler than the air in the summer and warmer in the winter. The earth's subsurface is an enormous heat sink—a solar battery—and it takes a large amount of energy to keep it in equilibrium. This heat energy comes in great part from the sun, a renewable and inexhaustible source of energy. In lesser amounts, it also comes from the center of the earth that we now know is a heat generator. The inner core of the earth is primarily made of a solid sphere of iron within a larger sphere of molten iron. Calculations show that the earth, originating from a molten state many billions of years ago, would have cooled and become completely solid without an energy input. It is now believed that the ultimate source of this energy is radioactive decay within the earth that continues to this day; the decay produces gradually diminishing temperatures from the earth's center to the surface. This does not mean that dangerous radioactivity is a hazard to us. We can tap into all of this heat energy, transfer it into our

How a Refrigerator Works

Like a refrigerator, a geothermal heat pump simply transfers heat from one place to another. When a refrigerator is operating, heat is carried away from the inside food-storage area to the outside area—your kitchen. Cold is not being added; heat is being taken out.

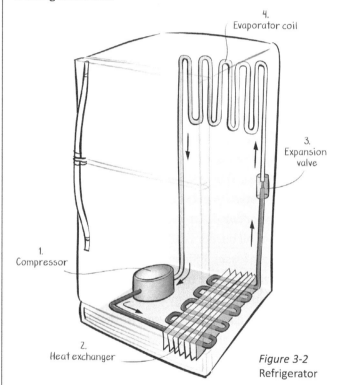

Figure 3-2
Refrigerator

To understand the operation of geothermal heat pump, it helps to understand how a refrigerator works. A refrigerator uses a refrigeration loop with four components:

1) A compressor
2) A heat exchanger
3) An expansion valve (often called a TXV, Thermostatic Expansion Valve)
4) An evaporator coil

A refrigerant is pumped through the loop to transfer heat energy from the inside to the outside.

The compressor (1) is a pump (basically a motorized piston) that pressurizes the refrigerant. Since temperature and pressure are directly related, as the pressure increases the temperature also increases in direct proportion.

The high temperature, high-pressure gas flows from the compressor to the heat exchanger (2) that is located outside the refrigerator. The heat energy in the heat exchanger is transferred to the cooler air in the kitchen and the refrigerant is condensed into a liquid.

Now the refrigerant is forced through a small orifice called thermostatic expansion valve or TXV (3). The small hole creates a pressure difference between the two sides of the device. This is like a dam on a river with a hole in the dam. Water leaking through the hole is at low pressure on the downstream side, but the water on the other side (being held back by the dam) is at high pressure. Again, the pressure/temperature relationship (low pressure/low temperature now) creates a cold, low-pressure liquid that flows through the evaporator coil. Relatively warm air from the food inside the refrigerator passes over the evaporator coil (4), giving up its heat energy to the cooler refrigerant. The refrigerant evaporates into a gas and the cycle starts all over again.

To make this happen, a geothermal system uses three loops to capture and transfer the earth's heat energy *(Figure 3-3)*. Each loop carries heat energy to the next loop. The **first loop** is a series of closed-loop pipes buried in the ground below the frost line in either a horizontal or vertical configuration, or placed in a lake or pond. Cool water from the unit in the house circulates through the ground pipes and the warmer earth releases heat energy into the water as it travels back to the GHP unit in the house. Non-toxic antifreeze is added to the ground loop water since it can sometimes be cooled below freezing by the GHP before the water circulates to the ground loop.

Or, as in my case, it is also possible to use an open-loop system

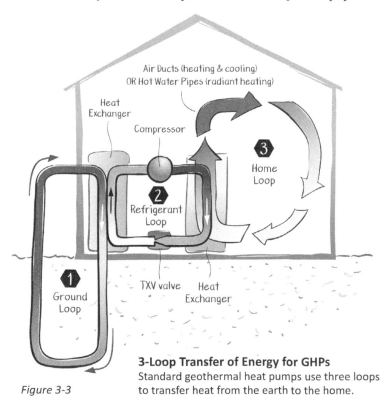

3-Loop Transfer of Energy for GHPs
Standard geothermal heat pumps use three loops
Figure 3-3   to transfer heat from the earth to the home.

by piping some of the well water directly through the GHP unit and returning it back to the well. There is no contamination of the well water; it merely circulates through plastic pipes before emptying into the well (more details in the next chapter).

The warmed water from the earth passes through a heat exchanger (coaxial copper pipes) to the **second loop** containing an even cooler liquid refrigerant. Heat transfer takes place which cools the incoming water and sends it back to the ground loop to pick up more heat energy. The refrigerant accepts the heat energy and becomes a gas as it heats up. The now gaseous refrigerant is sucked into a compressor where it is compressed and superheated to about 165°F (74°C). This large and important jump is discussed in more detail in chapter 17.

In a water-to-air system, the heated refrigerant then passes through a radiator-like heat exchanger (air coil) over which air is passed. Or in a water-to-water system, 120°F (49°C) refrigerant is passed on to a hot water heating system by means of a heat exchanger (hydronic coil) connected to baseboard registers or in-floor heating tubes. Air delivery ductwork of course, permits air conditioning. Again, chapter 17 provides a more technical description of how the gas laws and the three laws of thermodynamics come into play.

After the refrigerant transfers the heat to the air coil, it goes through a thermostatic expansion valve (TXV) and the pressure is released. The refrigerant becomes very cold (sometimes below freezing) as it circulates back to pick up more heat from the ground loop.

The **third "home" loop**: In a water-to-air system, the fan-driven cool air is heated as it passes over the air-coil heat exchanger and is then ducted into the home at about 105°F (41°C), transferring its heat energy to the walls, atmosphere, you, etc. As the room air cools, it is returned to the GHP to pick up more heat energy. In a water-to-water system, heat is transferred throughout the home through tubes in the floor or walls or baseboard registers.

How It Works: Cooling

Cooling is a similar process except that with just one tap of a button on the thermostat, a clever reversing valve sends the hot compressor output to the returning ground loop instead of to the indoor air loop.

While a geothermal heat pump operating in cooling mode uses essentially the same theory of operation as a residential AC system, there are two major differences.

First, a standard AC system requires a large outdoor heat exchanger in your backyard to dump heat energy into an already hot and, in some locales, humid atmosphere. This works, but is very inefficient because it takes a lot of electrical energy. The GHP needs no noisy outdoor box; all components are underground and inside the home. In my house, I send cold well water to flow over the heat exchanger pipes, heating the water which then flows back into the well, which in turn puts heat back into the ground.

Second, a standard AC system and a fuel burner/boiler are two separate systems connecting only at the ductwork. The GHP is a single system that can be reversed with a simple switch to provide cooling or heating. This reduces costs and increases efficiency. Alternately, we can set the thermostat for a given temperature and let the system determine if heating or cooling is needed.

That is it—a very simple concept, just three loops to bring in comfortable, quiet heat and cooling using well-established technology. The next chapter goes into more detail to describe different solutions to access that ground heat energy. ✪

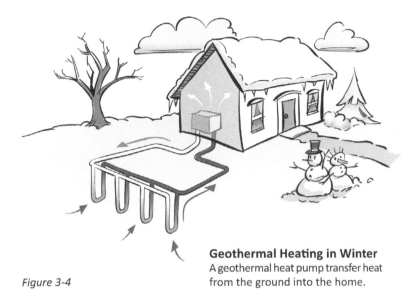

Figure 3-4

Geothermal Heating in Winter
A geothermal heat pump transfer heat
from the ground into the home.

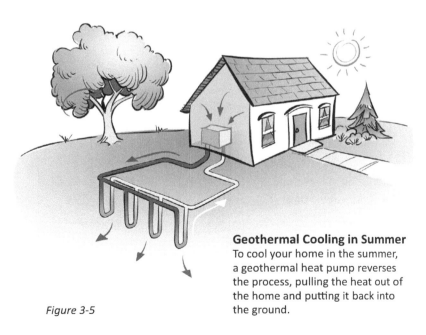

Figure 3-5

Geothermal Cooling in Summer
To cool your home in the summer,
a geothermal heat pump reverses
the process, pulling the heat out of
the home and putting it back into
the ground.

Water-to-Water GHP Systems

I have referred to water-to-air GHPs throughout this book, primarily because 75–85% of all units shipped in the US are installed in homes using forced-air heating/cooling. European and Asian markets, however, install 80–90% water-to-water systems for radiant heating, and they have several positive features.

Radiant water-to-water systems are physically smaller because they do not need bulky air-handling equipment. An added benefit: no ductwork means no heat losses in the ductwork.

In-floor radiant heat systems are quiet and invisible with no registers or baseboard units. People with allergies believe that radiant heat does not distribute allergens as much as convection-type systems which blow air around the rooms.

Water-to-water radiant heating is more efficient than water-to-air convection heating because the GHP can operate at lower temperatures. How are lower temperatures obtained? Simply by lowering the compressor pressure (since pressure and temperature are directly related). Less electricity is then needed to run the system, and that raises the coefficient of performance (COP), as explained later in the book.

Cooling is the biggest drawback with water-to-water systems because of condensation. Sending cool water through an entire radiant floor can produce serious problems. (Water-to-air systems can also produce condensation, but it is easily collected at a single point in the system.) However, water-to-water systems are being successfully installed in low-humidity climates, such as the southwestern United States.

Chapter 4

Ground-Source Solutions: Open vs. Closed Loops

The choices between the various methods of tapping into the ground for heat energy will in most cases be determined by the size of the home; the size of the GHP unit; the amount of available land; the type of ground (clay, wet, ledge); well capacity; availability of a pond, lake or ocean; number of degree days, and your budget. In fact, it is possible that there may not be many choices; there may be only one good solution. I was able to get an experienced geothermal contractor to sort this out, but it is always helpful when the homeowner is knowledgeable about the terminology and options.

The solutions for moving energy from the ground into a home fall into two categories, open loop or closed loop.

▸ In **open loops**, the water is not recirculated; it is just pumped back to the source.

▸ In **closed loops**, the same water is constantly recirculated within sealed pipes/tubes.

The piping for either system will be capable of transferring heat from the geothermal heat pump into hot water radiators or radiant heating (water-to-water), or to the air duct system (water-to-air). Some homes even have a hybrid of both setups.

Open-Loop Systems

Open-loop systems (sometimes referred to as pump-and-dump systems) are installed less frequently, but they can be the least expensive method if ground water or a high-capacity well is available. They are the simplest to install and have been used for decades where local codes permit. This type of system uses ground water from an aquifer or lake which is piped directly to the building's pressure tank and then through a water filter to the heat pump. The GHP manufacturer specifies the gallons per minute (gpm) or liters per minute (lpm) required. If you need a well capacity of 8 gpm (30 lpm) for a 4-ton GHP unit and 2 gpm (7.5 lpm) for domestic usage, then a 10-gpm (38 lpm) capacity well can provide both needs.

Once the water is cycled through the geothermal system, it is returned to the aquifer by discharging it through a properly sized drain field, river, lake, another well, or the same well. A water filter is required to keep out contaminants and must be cleaned regularly. City and county regulations may be imposed for this type of system.

....

Water quality is a major consideration for open-loop systems. If very low pH (< 6), very hard water (>100 ppm as $CaCO_2$), or hydrogen sulphide (rotten egg smell) characterizes the water, careful analysis should be done before using it in a heat pump.

....

Figure 4-1

Open-Loop Installation
An open-loop system uses ground water from an
aquifer (well) or pond which is piped directly to the
building's pressure tank and then to the heat pump.

Why Are Open Loops More Efficient?

It turns out that open-loop systems always have higher efficiencies than closed loops because water conducts heat better
than dirt. It's as simple as that. Closed-loop systems depend
on getting heat from sealed tubes that run through the earth.
Open-loop systems send water into the geothermal heat pump
so the heat can be extracted.

Winkler, Manitoba, Canada

Earth Smart Home

A modest Canadian bungalow of 2,080 square feet (193 sq m) including the finished basement, plus a two-car garage, sits on an exposed prairie near the town of Winkler, Manitoba. Their 5-ton, two-zone, horizontal closed-loop Northern Heat Pump system is a bit unusual in that it is multifunctional. First, in a water-to-air mode, the system delivers conventional forced air through ductwork for heating and cooling. At the same time, it has a water-to-water mode for radiant floor heating.

Radiant heat is a very comfortable approach because, for one thing, it is radiant instead of conductive. In addition, since heat rises, it is best to have the heat source as low as possible. The floor is as low as you can get. However, a radiant system is slow to heat and slow to cool because it is a slab of massive concrete. Now, it would be great if it were possible to insert chilled water into the floor in the summer to get room cooling via that floor—but, alas, it just won't work. When you cool air, it cannot hold as much moisture. The water vapor condenses out and becomes a problem. So a forced air system is needed for air conditioning. That also has condensation, but it is located at a single point so that it can be drained.

For Jacob Rogalsky, the issue is economy. He averages $85 (Canadian) a month in electric costs for heat, air conditioning and hot water all year. His neighbors with oil heat average twice that every month. "It's reliable. It beats the fuss about getting timely delivery of oil or propane when you live in the country. It's quiet. It doesn't take up much space. But the most important thing is that it saves so much money," Rogalsky says.

Photo courtesy of Northern Heat Pump

One of the newest approaches, **a Slinky™ coil technique**, is growing in popularity since it requires less trenching. A Slinky coil is flattened, overlapped plastic pipe in a circular coiled loop *(see photo in chapter 7)*. It concentrates the heat transfer surface into a smaller volume and thus reduces land area requirements by two-thirds. Large amounts of pipe are needed with this design. For example, one contractor provided 100 feet (30 m) of trench with 900 feet (274 m) of pipe per ton of capacity.

Figure 4-3

Closed-Loop Slinky Coil Installation
This method requires less trenching, but large amounts of ground-loop piping.

....

"Installation of the ground field is the most critical aspect of a GHP system, whether it is DX or water based." Curt Jungwirth, 35-year GHP expert
....

Up to this point, I have shown that horizontal closed loops have required trenches in order to insert the piping. But there is another closed-loop approach that is usually less costly and certainly a lot less messy.

Where there is adequate space for a horizontal loop but a desire to minimize disruption on the surface, the **trenchless horizontal bore loop** may be the preferred solution. Using special equipment that bores holes horizontally under the surface, the operator directs the machine to drill at a slight angle down to a typical depth of 10–20 feet (3–6 m). Using the right technique, he/she can "steer" the drill head to go deeper or shallower, or turn right or left. The drill head emits a radio signal that can be detected by a special device that tells the operator exactly where and how deep it is.

A small-diameter tunnel is created underground by displacing soil with pressurized water. The operator drills the horizontal bore, then directs the drill bit to come back to the surface, typically, about 200 feet (61 m) away. The drill bit is then removed and two ends of the loop pipe are attached to the drilling pipe and it is pulled back through the borehole, burying the piping underground. Finally, the borehole must be grouted.

This technique of horizontal directional drilling (HDD) allows the loop to be placed underneath homes, basements, driveways, wooded lots or even swimming pools. The only digging required is for the header unit and the supply/return piping into the house. This type of loop is especially attractive in a retrofit situation which minimizes disruption to the landscape. Please note, however, that clay soil may not be appropriate due to air gaps in the soil. And care must be taken when pulling piping through the hole since hitting sharp rocks can result in system leaks.

Not all contractors have this hole-boring equipment. To

Enfield, Connecticut

Earth Smart Home

This 2,100 square-foot (195 sq m) Cape Cod-style home in Connecticut had an inefficient, 33-year-old oil furnace that cost the homeowners $3,300 every year for heating and hot water. The owners, avid followers of alternative energy solutions, decided to go geothermal for both philosophical and economic reasons.

Terraclime Geothermal of Florence, Massachusetts, first conducted a detailed Load Calculation study using a "Manual J" software program. This is a common industry term that represents an analysis of the residence to determine the heating/cooling load. That, in turn, determines the size of the heat pump and then, finally, the configuration needed for the loop field. At one time this was a back-of-the-envelope calculation, but now experienced system contractors use a computer program to organize and compute the data. A contractor should be obligated to provide this analysis to the owner as a part of the bid. Terraclime not only did this, but they provided a comprehensive Energy Cost Analysis which compared their oil burner costs to the new heat pump installation costs.

The total installation, completed in 2008, took only five weeks. Their 3-ton DX system was made by Advanced Geothermal (ECR Industries) and six 70-foot (21-m) boreholes were used for the ground loop. The result: the GHP provided heat throughout that first winter without relying on the backup oil burner. Annual costs were reduced to $583 (for electricity to operate the GHP) and that included winter heat, full air conditioning, dehumidification, and partial hot water. In 2008 their Federal tax credit was limited to $2,000 (now increased to 30% with no limit) plus the State of Connecticut provided a $1,500 rebate.

The owners are pleased to report minimal maintenance, "All I have to do is change the air filters every six months—much less work than writing a check to the oil company each month."

locate a contractor in your area that uses this technique, do an online search for "geothermal horizontal drilling." Make certain that the driller has completed many such installations.

Vertical-closed loops are often needed if land surface is limited. Drilling equipment bores vertical small diameter holes 50 to 600 feet (15–183 m) in depth to hold the plastic piping. This is usually the most expensive approach because of the drilling expense. However, it is very useful when the land is rocky. The number of holes and depth varies depending on many factors, including soil and rock condition. Some locations require 180 feet (55 m) of borehole per ton and 360 feet (110 m) of piping. The closed loops contain water or a mixture of water and antifreeze. A special grout is often pumped down the holes to surround the pipes from bottom to the top to insure good heat transfer from the surrounding earth. ✪

....

The choices between the various methods of tapping into the ground for heat energy will, in most cases, be determined by the size of the home; the size of the GHP unit; the amount of available land; the type of ground (clay, wet, ledge); well capacity; availability of a lake or ocean; number of degree days, and your budget.

....

Figure 4-4

**Closed-Loop
Vertical Installation**
Multiple wells are drilled
for the plastic piping. The
depth will be determined by your
soil type and size of heat pump.

West-Central Minnesota

Earth Smart Home

I remember Minnesota. It is a great place with good people. I also remember having my car tires freeze flat on the bottom, causing a terrible thump/thump/thump. It does get cold there. If a GHP system works there, it will work anywhere.

To prove the point, I want to describe a West-Central Minnesota residence that proves that a GHP system installation can be flexible enough to adapt to local topography and individual architecture.

Built in 2004 for a family of six, this home has 4,300 square feet (399 sq m) of space on two floors plus a basement and garage adding another 864 square feet (80 sq m). They installed three GHPs, with a total 10 tons of capacity. The manufacturer was ECONAR, a Minnesota manufacturer and the same company that supplied my unit.

The ground source was planned to be 150- to 165-foot (46–50 m) deep boreholes but they hit bedrock at 135 feet (41 m). It was better to dig more holes than dig deeper, so they ended up with 13 boreholes plus a well for water.

They have three zones: 1) Top floor: water-to-air to provide heat and AC; 2) Main floor: water-to-air to provide heat and AC; 3) Basement and garage: water-to-water for radiant heat.

A unique feature is that they installed one of the GHPs on the top floor since it is more efficient to create the cool air closer to where it is needed.

The total cost of $39,000 included the boreholes, ductwork, a backup gas furnace (required by the local utility co-op in order to get off-peak electric rates), radiant heat in the basement and garage floor, a desuperheater, a backup water heater, and a Heat Recovery Ventilator (this permits a tighter home to save additional costs while still circulating fresh air in every room).

If they had used a standard propane heating system, it would've cost approximately $5,400 per year (at $2/gal for LP). The average annual electrical cost just to run the GHP units is about $1,200. So they are saving close to $4,200 every single year. At the same time, their home is very comfortable in every season and with the desuperheater, they have free hot water in summer (and almost free in the winter) for six people. They are especially happy with the good, consistent humidity and the even distribution of heat and cooling. This was a very successful installation that is the result of selecting a highly qualified installer with over 15 years of experience.

Chapter 5

Direct Exchange (DX) Systems

The DX, Direct eXchange concept (sometimes referred to as DGX, Direct Ground eXchange) is a closed-loop system which uses two loops instead of three. While it may not be the least expensive in terms of installation nor the most popular with installers, comparative testing has shown a 20–30% increase in efficiency with DX systems.

In the preceding chapters, I have described the three loops used to transfer heat energy from the ground into a home. The DX system combines two of those loops by using a refrigerant in the ground loop. This elimination of a loop and its related equipment has immediate advantages:

▸ Increased system efficiency due to one less heat-transfer loop and the use of high-pressure refrigerants.
▸ DX systems require more refrigerant to fill the underground copper pipes.
▸ Eliminates the need for a water circulation pump since the compressor does all the work.
▸ The higher conductivity of copper pipe reduces excavation and installation costs (smaller diameter and shorter lengths of pipe and fewer, shorter boreholes).

The US EPA Energy Star Heat Pump criteria set a minimum efficiency for DX systems at 350%. DX manufacturers claim performances ratings from 400% to 580%, depending on the configuration. Without a water circulation pump, the "cost" of circulation is transferred to the compressor, and this is reflected in slightly lower COP ratings. (Energy usage of water pumps are not included in standard 3-loop GHP efficiency ratings.)

When DX systems were first introduced in the 1970s, there were some failures due to poor installation procedures. In those early DX systems with vertical boreholes and very long, high-pressure refrigerant loops, some of the oil from the compressor moved into the refrigerant, thus reducing the lubricating capabil-

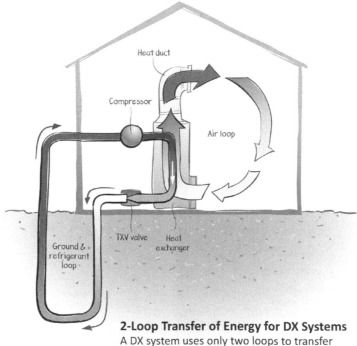

2-Loop Transfer of Energy for DX Systems
A DX system uses only two loops to transfer energy; although often not the least expensive system to install, it can achieve higher efficiencies.

Figure 5-1

ity of the compressor and damaging it. In addition, copper pipe fractures in the ground loop caused by ground heaves (due to the earth being frozen) allowed refrigerant to escape into the soil and ground water, posing a threat to the environment. These were serious and expensive problems for the homeowners.

The technology and installation techniques have greatly improved, and while not the most popular, large numbers of DX systems are in operation in the United States and Canada.

DX System Considerations

▸ A less invasive approach to loop fields utilizes 30-degree angled shafts, each less than 3 or 4 inches in diameter.

▸ Special brazed copper is now used for tubes that are adequately spaced apart. Pressure testing is a must.

▸ The empty space around the copper pipe is filled with heat conductive, cement-like protective grout.

▸ Oil separators or other solutions are now a part of all DX systems to insure proper oil return to the compressor.

▸ Although ground acidity is not common, DX installers must measure the soil pH where the manifolds will be located. If the pH is 6 or less, the loops need cathodic corrosion protection (much like a boat propeller in seawater). If not, the copper pipe will deteriorate over time. Installing DX systems in limestone is not a good idea.

▸ Although special techniques are required to drill in wet areas, the results can be worth it due to better heat transfer.

Professional installation for any geothermal system is critical, but even more so in a DX installation. Leakage, cavitation and pump failures can still happen from improper installation. If a leak should occur on the DX earth-loop side, the repairs and refrigerant recharging would be a substantial expense. Check the warranty on the installation, not just on the unit itself. Work only with an experienced contractor. ✪

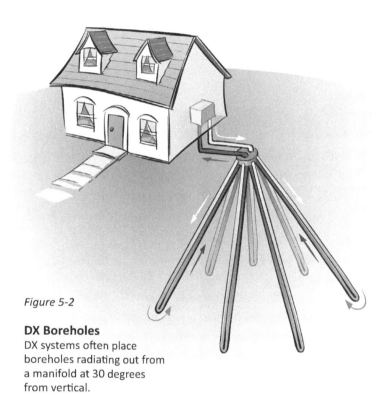

Figure 5-2

DX Boreholes
DX systems often place
boreholes radiating out from
a manifold at 30 degrees
from vertical.

By using underground thermal sensors, Sub Terra *(www.sub-terra.com)* has determined that most of the heat transfer takes place within a 2-foot radius around each DX tube. This **area of influence** collectively becomes a reservoir of heat capacity. For their DX installations, they keep at least 4 feet (1.2 m) between tubes to avoid robbing heat from one another tube and to prevent the ground from freezing.

Cuddebackville, New York

Earth Smart Home

This stone house in Cuddebackville, New York, is unique for several reasons. Built in 1753 when King George of England ruled the colony, it was converted to a home in the 1930s with an oil-fired steam radiator. Now it could certainly be the oldest home in the US retrofitted with a DX geothermal system.

Because of its age and construction, special consideration had to be given to the building before any heating installation was started. After an evaluation by a Certified Building Professional, Total Green Geothermal of Monroe, New York, did a GHP software study to determine the heating requirements, how best to distribute the heat, and how to create a "thermal envelope" that included insulation of below-grade walls in the basement, high-density fiberglass in the attic, and air sealing the living space. Normally, a GHP delivers quiet, low velocity warm air to the living spaces. But, in this case, to retain the unique character of the building, a special smaller, high-velocity ductwork was chosen.

Heating bills prior to the installation were over $5,000 per year. Now it costs $65/month for heat, hot water and full air conditioning. For a musician, this is especially important in order to protect sensitive musical instruments. The owner says, "I am going to recoup my investment in seven years. Next, I plan to install solar panels." The 5-ton DX unit was built by Advanced Geothermal Technologies in Reading, Pennsylvania. Ten holes, each 70 feet (21 m) in depth, were drilled at a 30-degree angle.

It is good to see an older building in such excellent use today.

Photo courtesy of Total Green Geothermal, LLC

Chapter 6

Using a Well as the Heat Source

I was intrigued the day my geothermal contractor suggested using our well to transfer heat to our proposed geothermal heating system for our-yet-to-be built home. Our new well had a capacity of 12 gallons per minute (45 lpm) and it seemed to be the perfect solution.

After almost four years of use, I consider it a very good approach for our situation. But like most things, every approach has its tradeoffs. The major advantage of an open-loop well-source system: lowest installation cost. Major disadvantage: higher electrical cost to run and maintaining a water filter.

I recently read an article about a homeowner who had installed a well-based geothermal system in an existing home. Naturally, I could not wait to read it in order to compare ideas. He was very enthusiastic about his system, but had real problems getting it to work. An important lesson can be taken from that article. First, he clearly chose a contractor with no experience in the installation of a well-based system, costing him discomfort, time, and money. One thing the contractor missed was the fact that in conditions of very prolonged cold weather, the heat taken out of the well can, at times, cool down the water and make the system less efficient.

When we use ground trenches or deep holes for heat, we are tapping into an almost infinite heat sink. But an aquifer is smaller and more subject to changes. My well-water temperature runs about 50°F (10°C) in the summer as I add the heat extracted from the cooling of our home, and about 43°F (6°C) in the winter as I transfer heat energy out of the well.

Both my geothermal contractor and well contractor recognized that seasonal temperature changes in the well water (cooling in winter; warming in the summer) could be problem and planned a solution that was implemented from the start. They installed a third pipe from the GHP, through the basement, past the well and toward a stream in the rear of the house. When the well water temperature reaches 40°F (4°C), the system automatically diverts the really cold water to the third pipe. It only runs in that mode for an hour or so and then switches back. It is so quiet that I never know it is happening unless I am passing near the outlet. It is double piped for weather protection and has a grid at the end to keep out wildlife.

There are some who are concerned about the impact of a well-based system on the aquifer. Potable underground water is a precious resource. The use of plastic pipes in place of copper is commonplace now for very good reasons. But there is concern that they may be adding pollutants to the water, especially if large numbers of homes are involved. My research shows that this concern is probably unwarranted.

The materials used are proven to be safe: cPVC (chlorinated polyvinyl), HDPE (high-density polyethylene) and PEX-a and PEX-b (which are "cross linked" polyethylenes) are commonly used water piping. Each have been approved and certified for use in drinking water applications by NSF International and ANSI, the American National Standards Institute. However, scientists at the University of Virginia have reported to the American Chemical Society that their studies have shown that some plastic pipes can affect the odor and/or taste of drinking water: HDPE had the highest odor production, but did not release organic materials; cPVC and PEX-a have a low odor potential and did not release organic chemicals; PEX-b had a moderate release of odors and materials. All of this seems to disappear in a month or two of use. There is no data that shows any adverse health effects because of these materials.

I had a few more thoughts on this subject as I made my temperature rounds on a cold winter morning. It was our coldest outside air temperature that winter at 1°F and a wind chill at -8°F (-22°C). Inside it was a comfortable 70°F (21°C). We like to keep the night setback at 68°F and set the system to automatically move to 70°F at 6:00 a.m. I measured the timing at the changeover—it took less than five minutes. This told me that the system was working well—in spite of the cold. Then, in washing up, I was again struck by how really cold the water from the bathroom faucet feels at this time of year—43 degrees actually.

Well Source Advantages/Disadvantages

Using a well as a heat source is usually less costly than the trench or deep-hole approach. It is also lower in cost than a traditional furnace and full-home air conditioning. This means that an immediate positive payback is gained because of the savings in oil costs and construction costs. Other aspects of using a well:

▸ Requires minimal maintenance; just clean the water filter 6 times a year.
▸ The process does not pollute the water.

▸ A well-based system uses more electricity in the winter months because the submersible well pump is running quite often. For my system, this amounts to about $100–$150 per month for four months. Since we only use summer air conditioning in very hot, humid weather, we rarely have additional electrical costs during those times. The net gain from the elimination of oil costs and oil furnace maintenance is far above this.

▸ The well temperature can be lowered during long periods of very cold weather or raised in long periods of very hot weather, requiring an alternate disposal pipe.

▸ Very dirty or contaminated water should not be used.

▸ Local codes/restrictions may apply. ✿

According to data supplied by the US Department of Energy Office of Geothermal Technologies, nearly 40% of all U.S. emissions of carbon dioxide are the result of traditional heating, cooling, and hot water systems in residential and commercial buildings. This is roughly equivalent to the amount of carbon dioxide contributed by automobiles and public transportation.

Chapter 7

Geothermal
Advantages / Disadvantages

Of the dozen advantages listed on the next page, the driving force behind my decision was the first one—the urgent desire to eliminate fuel-oil costs. The rest are very real and very much appreciated, but they really come with the package.

We cannot ignore, however, the disadvantages listed below. How important is a higher installation cost? If a residential GHP costs more than a standard heating and cooling system and you plan to move in the next three to five years, why bother with a GHP? Well, since home heating/cooling can be an expensive proposition, you would still profit from years of zero fuel bills. And if the extra cost difference is added to your mortgage, the added annual mortgage costs will be less than the annual cost

savings. The payback is immediate. Then factor in the 30% federal tax incentive, and a possible state or province incentive. In addition, the increase in your home's value could more than pay for the difference.

GHP Advantages

▸ Fuel costs (oil, natural gas, propane) are zero, year after year.

▸ GHP systems conserve natural resources by using the earth's renewable energy.

▸ GHP systems provide heating *and* cooling and can change modes with a flick of the thermostat.

▸ GHP systems have lower maintenance costs, no oil filters or nozzles to clean, no open flames, no carbon monoxide, no chimney, no air pollution, no oil storage tanks, and no sludge.

▸ GHP systems are by far the most efficient (400–500%) system available in converting a small amount of electrical energy into a large amount of home heating/cooling.

▸ GHPs have a notably quiet operation.

▸ When the system is running, a desuperheater water heater heats domestic water, saving additional costs.

▸ Long system life: you can expect 20–25 years of service, which is double that of a boiler or furnace and AC system. The plastic loop pipe is guaranteed for 50 years.

Slinky coils for a GHP system prior to burial.
Photo courtesy of GeoSystems, LLC

▶ Increased home value (see the next chapter).

▶ More affordable now thanks to federal subsidies, tax credits and rebates (see chapter 12 for details).

▶ There is a measure of satisfaction in watching the fuel delivery trucks pass by without stopping. And you are granted bragging rights (or at least a dinner party conversational topic) about how you heat your home without fossil fuels.

GHP Disadvantages

▶ Most closed-loop geothermal installations cost more than a fuel burner plus full AC, perhaps 20–30% more. This is because of the drilling or earth excavation and the installation of the pipe for the vertical or horizontal ground loops. However, an open-loop heat source can often cost less than a fossil-fuel system plus full AC.

▶ While a GHP is simple in concept, compared to traditional heating and cooling systems, it is more complex to install. It has advanced electronics, sensors, control loops, and safety controls, which means the installation and service personnel require a higher level of training. It's like the difference between maintaining your 1955 Ford by yourself as opposed the professional skill required to maintain a new SUV. See chapter 15 for more information on this topic.

▶ While you are saving on fuel costs, you may experience an increase in electricity usage, depending on your configuration. Remember, though, a small amount of electrical energy adds a large amount of heat energy. ❂

The EPA has stated that geothermal systems increase the value of a home $20 for every dollar saved in oil/gas heating costs per year.

Chapter 8

Does a Geothermal System Really Increase a Home's Market Value?

My original premise that a ground-based GHP will provide an increased market price was simple. It seemed intuitively logical that if two homes were equal except that one had a GHP system, the GHP home would be more desirable and a buyer would pay more. Then I found that the EPA had written the same thing more explicitly. They stated that the home's value increases by $20 for every dollar saved per year in heating costs.

So if you would have used, say, 1,000 gallons of heating oil at $2.50 per gallon, a GHP system would save $2,500 dollars per year and that would increase the home's sale price by $50,000. This is not unreasonable. If a buyer added the increased cost as a part of a 30-year mortgage, the annual increase in payments would be less than the annual savings in fuel costs.

But the EPA does not explain how they came to this conclusion. And you could argue that the EPA does not establish market values—the marketplace does this. So I have wondered how realistic is this? How could anyone verify, or quantify this? Well, I could sell my home. That would give me one data point. But I'm not that crazy. Could I spend a scarce $375 to have a certified appraiser establish my market value? I could and I did.

Let me review how appraisers work. Ideally, they examine the database records to find three homes in your area that have sold over the past year, and are exactly like yours. Then they average the prices and presto—they have a market value. Of course, they will not find even one home just like yours or mine. In fact there are no homes with geothermal heating systems that have been sold in our area. So the appraiser finds at least three homes that are generally in the assumed price range, and sets up a series of increases and decreases in value to bring the homes into an assumed and approximate equality. Then that average would be a basis for the value of the appraised home. For example, if the comparable home had a two-car garage and mine did not, the appraiser could *lower* the value of the comp home by the value of a two-car garage. If I had a geothermal system and the comp house did not, the appraiser could *add* a value to the comp home. The question is—how much?

Here is what happened. Arthur was blindly chosen from a list provided by a mortgage company as a reliable, certified appraiser. He arrived and quickly started to sketch a floor plan using a laser rangefinder to get the dimensions so that he could add up the square footage. He then took interior and exterior photos with his digital camera. I gave him a basement tour showing the geothermal system quietly humming along (no chimney, no oil tanks). I showed him the water heater with its domestic cold water pipes "in" and hot water "out" on top of the unit and the geothermal hot water "in" and cooler water "out" at the base.

I explained how I was saving at least 1,000 gallons (3,785 liters) of oil per season, and quoted him the EPA statement about the increase in value because of a GHP. He listened carefully to my pitch, and then said that there would be some people who would not want a GHP because it was too different—just what I did not want to hear. He did state that FHA guidelines permitted an increase due to energy-efficient systems. After he drove off, I was left with much uncertainty about how this would turn out.

It took a week before I received the details in writing. Before I show the results, I must explain two things. First, during the negotiations for a construction loan, my bank required a certified appraisal based on the actual price I paid for the land plus the architectural drawings for the home. This gave me a baseline. Next, during the past year of occupancy, the value of US homes declined. In our county, it was small, perhaps 5%, but there was a decline in value.

The new appraisal showed a 14% increase in value! (I would not want the town assessor to read this.) That is about a 4.5% increase per year compounded. And in reference to the EPA estimate, my value increased by $25 for every dollar I saved in oil costs, even after I reduced those savings by the increased cost of electricity to run the GHP.

Appraisers are not buyers. It is impossible to know how important it is to a buyer to be free of fuel costs or whether the buyer understands that a GHP system is not "strange"—that it is basically established refrigerator technology.

What have I learned from all of this? I believe that a GHP system does increase the home value. How much of an increase will be unknown until there is an actual sale. In the meantime, we who have such systems have a good thing going for us. We are not about to sell. That tells us all something. ✪

Chapter 9

Learning to Live with a GHP System

Maintenance

You have a new GHP installation in your basement; a winter nor'easter storm is coming on. The system is so quiet that you have to keep checking to see that it is working. You are a bit uncertain and nervous, but you finally relax. Suddenly, the system quits! There is no heat. Your family looks at you. *Do something!* This was supposed to be so reliable! Doubts creep in. Have you made a huge mistake? Maybe a conventional furnace would've been better after all. Don't panic—call your contractor.

This happened to me three times. In every single case, the GHP system did exactly what it was supposed to do—it shut itself down because of preventable maintenance problems.

The first time was because of a dirty air filter. When we were in the drywall construction phase, my builder did not want to leave an unattended propane heater on all night for safety reasons. But it was freezing weather and sanding and plastering were underway. The GHP had been installed, but the well water was not yet hooked up. So I decided to use the electrical backup capability to provide heat. This worked fine, but even though the return ducts were disconnected, a great deal of dust got into the system, clogging the new air filter. Lesson learned: never turn on a GHP during construction. Clean the air filter regularly and record it on the maintenance sheet. (I later learned dust could also get into a heat exchanger, which would void the warranty.)

The second shutdown occurred because a mouse chewed through the well pump control-wiring insulation, automatically turning off the GHP. Mousetraps are now checked often.

The third time it was the water filter. Any new well often has particulate matter. It can easily be cleaned, as I found out. Set up a schedule and record it.

From that point on, the system has worked flawlessly, mousetraps included. By the way, I discovered that better mousetraps *have* been built.

There is one other lesson to be learned. If you need to clean the filters while the system is on, you obviously must turn the system off first. When you finish cleaning and turn it back on, you may not realize that there is a time delay of about five minutes to protect the compressor. Even after that it may not start. An age-old trick works every time—turn the GHP circuit breaker off and then back on. Get to know your circuit breaker box.

Backup Systems

In colder climates, GHP systems always include a built-in electrical backup heater that can take over in case of, say, an unlikely compressor failure or even as added energy in case of very, very cold weather. But the contingency that concerned me was a power failure in the middle of a serious snowstorm. Propane-fired emergency power generators are a neat answer, but expensive for me. I picked a wood burning stove as my backup. So, I may be burning something after all!

Night Setbacks

With an oil burner, we assume that we can save energy costs
by turning down the thermostat three to five degrees at night,
depending on our comfort level. But with a GHP, I noticed that
the second-stage compressor *(see next chapter for more info on
stages)* would often kick in, adding costs and creating doubts
about a setback. So I checked with my manufacturer who con-
sulted with the thermostat manufacturer. The answer came
back: "The more efficient the system, the less savings with night
setbacks. The most efficient approach with a geothermal system
is to set the thermostat and never change it." But I preset it back
at night for comfort, not savings. Of course, if the home is going
to be unoccupied for long periods, use a setback for savings.

Comfort Level

In our former home, the schoolhouse, the oil furnaces were in the
basement just under our bedroom. We actually had three smaller
furnaces—for the studio, first floor and second floor—instead of
one large one. Every time that thermostat kicked on a furnace,
we heard a loud roar as a sudden rush of air and combustion
took place. We seemed to have problems with incomplete combustion because the nozzles often became clogged with residue
from partially burned oil. And the transformer located near the
combustion chamber kept burning out. We needed an annual
cleanup to maintain efficiency (it was certified as 86%, but that
soon deteriorated). We needed 3,100 gallons of fuel oil each year
even though we set the second floor temperatures low.

In contrast, our GHP is quiet. At times I can just hear the
compressor running when it goes into stage two if I am in our
dining area directly above the GHP unit. In addition to the quiet
operation, the rooms are always comfortably warm in the winter.

We do not use AC too often, but being close to the Hudson River, it does get warm and humid. When we need AC, it feels so good! There is also comfort in not having to worry about an on-site supply of fuel oil getting low or running empty. And finally, the absence of flames and pollutants is comforting because it reduces the odds of something bad happening.

Thermostats

I am somewhat nostalgic for the famous Honeywell Round thermostat. I think every home we've ever owned had at least one, including the schoolhouse, which had three. The bimetallic scroll worked without any integrated circuit; it gave us the current temperature, allowed us to set the temperature and turn it off or on. It was simple, low cost, and doomed. That's because the on/off part was accomplished with a mercury switch—no longer a smart idea.

GHP thermostats are all digital. They hook into a digital circuit board and have more functions, though if you look at the outside case of a typical model, it seems simple enough.

You see just two buttons: they set the desired Set temperature Up or Down. On the left side of the display you see the Time which alternately flashes with the current Inside Temperature. In the middle is an AM/PM indicator. On the

Digital Thermostat

right is the Set Temperature. Below that is a Flame Symbol that indicates it is in the heating mode. And above the time is the Day of the week (WE for Wednesday). Not shown since the system is Off in the example: in the upper right corner would be an STG-1 or STG1-2 to indicate if only the first stage or both compressor stages are on.

Now, open the case. Six more buttons are exposed. It may look complicated, but the homeowner seldom needs to use all of these settings since most are only for the installer. This is a good reason why it would be smart to save your installation manual. Some directions are written on the inside of the cover, but you really need to follow your own specific thermostat directions.

All thermostats are not exactly the same, but will have similar features. Here are the buttons for this particular thermostat:

When first turning on the system, you will want to correct the clock setting using the Time button ③ and learn how to correct for Daylight Savings.

Then you will need the System button ⑧ to set Cool or Heat, Emerg or Off. (Emerg= Emergency is used if the compressor will not work and you want to use the backup electrical strip.) We

① Raises temperature setting

② Lowers temperature setting

③ Time button (changes thermostat Time, up or down)

④ Program button (sets thermostat from factory presets to your desired settings)

⑤ Run (run program) button

⑥ Hold temperature button

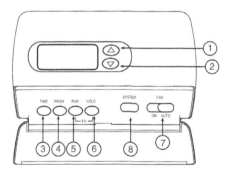

⑦ Fan switch (ON, AUTO)

⑧ System button (Cool, Auto, Heat, Emerg, Off)

Figure 9-1   **Digital Thermostat Schematic**

use the system button for comfort, knowing full well that it is more efficient not to adjust the temperature often.

Next, you can program the system for automatic temperature settings for day or night (refer to the manual). Here is where you set a heating/cooling schedule plan so that weekdays, Saturdays, and Sundays have a different schedules (or the same) and settings can vary throughout the day. You can always override the Program ④.

Still hidden behind the upper case are two AA cells for Time backup during power failure. The GHP will not work in a power failure, but the clock will. ✪

A typical geothermal heating and cooling system that also provides free domestic hot water. Photo courtesy of GeoSystems, LLC

Chapter 10

Reducing Operating (Electrical) Costs

Because they are more efficient, geothermal systems require less energy than standard HVAC/R system, but they still require electrical energy to work. The whole concept is to use a small amount of electrical energy to control a large amount of heat energy. So anything we can do to reduce electric power will increase system efficiency and cut monthly bills. The colder or hotter the weather, the more important this becomes.

Constant Pressure Controller

Since I am using the well as a heat source, my submersible pump replaces the electrical power that would be used to pump water through a ground loop. Our submersible pump is larger with multiple needs, and so it requires more power. The well contractor suggested a special control system that provides just enough power to meet the demand (instead of having the pump either completely on or off), while at the same time, keeping a constant water pressure. If I only use the shower, it provides enough power to provide water for that. If I have the GHP heat, clothes washer, and dishwasher all running at the same time, it will increase the power accordingly. It works in conjunction with a small pressure tank and a microprocessor to send the right amount of electricity to the pump to speed it up when more water is needed, or to slow it down when less is needed, while keeping the pressure at a constant 50 psi. Submersible pumps are now available that provide capability built-in.

Multiple-Stage Compressor

My GHP, like most units, has two levels, or stages, of compressor operation. A few systems have three stages. The GHP automatically decides when to change stages, based in part on the difference between the current temperature and the set temperature. Stage One is normal operation. Stage Two, which requires more power, goes into effect if the weather is particularly cold for a long period, for example. My goal is to keep the system in Stage One as much as possible. I am careful to only make small changes in settings at any one time in either heating or cooling mode. Unlike a standard compressor which only operates at full capacity, two-speed compressors allow heat pumps to operate close to the heating or cooling capacity needed at any particular moment. This saves large amounts of electrical energy and reduces compressor wear.

Scroll Compressor

Another advance in heat pump technology is the scroll compressor which replaces the piston design. It consists of two spiral-shaped scrolls; one remains stationary, while the other orbits around it, compressing the refrigerant by forcing it into increasingly smaller areas. Compared to the typical piston compressor, scroll compressors have a longer operating life, are quieter, and provide 10–15 degrees higher temperatures. A scroll compressor has a digital controller and fewer moving parts (only 6, compared to a piston type with 20 parts). It is more efficient, thus saving electricity.

Heating Domestic Water with a Desuperheater

Most geothermal units have an optional feature called a desuperheater, which heats domestic water for showers, etc. This component consists of a refrigerant-to-water heat exchanger installed at the discharge of the compressor. The hot gas at this point is in a "superheated" condition. In the desuperheater, the refrigerant releases some of the heat into the cooler water through the copper walls of the desuperheater heat exchanger. A small pump circulates the water from the electric water heater/storage tank to the heat exchanger, where it picks up heat energy and returns it back to the storage tank.

This excess heat energy is available in both the heating and cooling modes. However, there is a greater benefit in cooling mode since there is more heat to be eliminated. The amount of hot water to be generated is a function of the run time of the GHP unit. On very hot or very cold days the desuperheater may be able to generate the majority of the domestic hot water needs because of the longer run time. A safety cutoff switch shuts off the circulator if the temperature exceeds 130°F (54°C).

The desuperheater can reduce your water heating bill by about 20–40%. By design, it provides only partial water heating and you will still need a water heater. In the dead of winter when both the desuperheater and the heat pump are working, the amount of energy given to the hot water tank is fairly small.

Desuperheater capacity is directly related to GHP capacity, so the hot water provided by a 5-ton unit would greater than a 3-ton unit. It is far preferable to have your GHP installer also install the water tank since they must be appropriately linked.

A word of advice: watch for scale buildup in the desuperheater exchanger that could affect its performance.

Other Water Heating Options

If you have an electric or gas water heater you know that it is energy hungry. It is expensive to keep the tank at 120°F (49°C) even when no water is drawn. I chose a desuperheater to reduce these costs; it uses spare heat from the heat pump to heat my water. But you should be aware of other solutions.

Solar Heating: Solar water heaters can be a great bargain for heating domestic water if your home has good south-facing exposure. They require a storage tank (various sizes are available), but with longer winter nights and cloudy days, a gas or electric auxiliary booster is needed. More details about the various system options can be found in the book, *Got Sun? Go Solar.* A local installer can provide an estimate and federal tax credits are available.

Tankless Heating: On-demand tankless (or instantaneous) water heaters produce hot water only as needed, eliminating

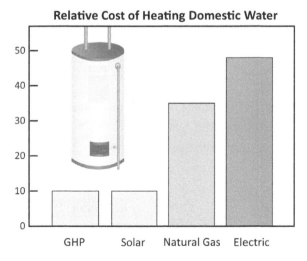

Figure 10-1 Solar and GHP heating of domestic hot water costs much less than natural gas and electric. Source: VDI (Association of German Engineers)

storage tanks with their standby losses and operating costs. Hot water never runs out unless the demand is greater than the designed flow rate (2–9 gpm; 7.5–34 lpm). These are independent of the home heating system and produce a constant supply of hot water on demand by using "flash" heating elements to reducing operating costs by 10–15%. They are the primary source and need no auxiliary backup. Costs are $800–$2,000 (plus installation).

Air Source Heat Pump: Another approach is to use a small dedicated, stand-alone air source heat pump, with its own compressor that uses the air in the basement as a heat source solely for domestic hot water. A storage tank is required. They cost around $1,500–$2,000 (plus installation).

On-Demand (OD) Water Heating: A few companies have integrated "on demand" domestic water heating into their GHP systems. ClimateMaster, for example, includes a dedicated hot water system that includes a "Smart Tank" to provide hot water even when the geothermal heat pump is not turned on, and it can do this five times more efficiently than an electric water heater.

While an OD system and a desuperheater are both integral to a GHP, it is interesting to see the differences between them:

▸ A GHP with a **desuperheater** gives "Home Heat Priority," only providing *excess* heat energy for domestic hot water. It does this only when the unit is providing heating/cooling to the home. A desuperheater requires a primary, separate water tank heated by electricity or gas. As an auxiliary system, it provides 20–40% of the hot water. The added cost is about $500.

▸ An **OD system** takes a "Hot Water Priority" approach. If the GHP is not in a heating or cooling mode, or is turned off, the OD turns on the compressor and uses it only for hot water. If the GHP system is running and there is a demand for hot water, the OD diverts heat energy for hot water. Its fast response has virtually no impact on home heating. It is a primary domestic hot water source with its own storage tank and does not require electricity or gas auxiliary backup. The OD provides 100% of domestic hot water needs. In 2010 it added $2,250 to the system cost. ✿

Swimming Pool Heating

I assumed originally that this might be a natural additional use for a GHP installed in a home since in the summer there is excess heat being transferred out of the home and into the ground. Why not send it to the pool instead? There are such installations that work very well. But the AC is not always turned on in the summer and it turns out that a separate heat pump is often needed for the pool. If you are interested, look for companies who provide dedicated heat-pump pool heaters/coolers. Federal tax credits do *not* apply.

One of the best pool heaters I ever saw was in the French countryside. It used black plastic pipes to trap the heat from sunlight and a small circulating pump. If you have ever tried to use your garden hose after it has been sitting in the sun, you'll understand.

Driveway Snow Melting

Keeping a driveway free of snow may be practical in some commercial applications where there is a great amount of refrigeration equipment that continually produces waste heat. But for residential use, it may require a heat pump larger than that needed for an average home. For more information, see "An Informational Survival Kit" at *www.HeatSpring.com*.

Chapter 11

Geothermal Plus Solar Energy: A Perfect Match

I f there were ever a marriage made in heaven, it would be to pair a geothermal heat pump with a solar-electric installation. They both use nature as an energy source. The GHP may be free of fuel costs, but it does need electricity to run. If you are connected to the electric grid, a solar photovoltaic (PV) system can reduce your electrical load because every watt-hour your solar system delivers is a watt-hour you do not have to buy from your utility company. And this is true whenever the sun is shining because sunlight is converted into free electrical energy.

The beauty of a PV installation is that when it is connected to the electrical grid, the grid is like an infinite, free storage battery. Solar power cannot only feed electrical energy into your home to reduce the amount needed from the electric company, but during those times when the solar system generates more power than you are consuming, it sells electric power back to the utility. It does this by automatically reversing the electric meter, spinning it backwards as the excess power is diverted to the grid.

If your utility offers net metering, your electric bill will be lowered and you will get a "credit" at the same rate you pay them. Not all states have laws for net metering although many utilities provide it anyway. One alternative is to use a second meter to measure the power going to the utility, but then you usually get reimbursed at a much lower rate. The Database of State Incentives for Renewable Energy's website *(www.dsireusa. org)* lists a wealth of information, including which states mandate net metering.

There are only three primary components that are wired together in a grid-tie system: an array of silicon photovoltaic solar panels mounted on an aluminum structure(s), an inverter to convert the direct current (DC) out of the array to alternating current (AC) that is usable in the home, and an outside 'reversible' meter supplied by the utility company. (Leased systems will also have an inside meter supplied by solar company to record the solar kWh supplied.) To learn more about your grid-tie solar options, read *Got Sun? Go Solar, 2nd Edition.*

With this setup, if the grid has a power failure, you have a power failure. A propane-fired emergency power generator is one solution. Battery backup for selected circuits is another.

Roof-mounted solar arrays are ideal for homes with south facing roofs that do not have large trees that would cause shading. PV panels can be mounted on flat or pitched roofs. Ground- or pole-mounted systems are also effective installation alternatives.

A typical small home may have a 450 square-foot (42 sq m) south-facing roof installation that can fit 30 PV panels that would, depending on climate and latitude, average 16 kilowatts-hours (kWh) per day of DC power. This would typically provide 6,000 kWh over a year; at $.15/kWh, that equals about $900 in savings annually. In my case, 6,000 kilowatt-hours would cover my geothermal power needs, but I would need a far bigger solar system to cover my total annual costs. Some analysts assume that electrical costs will increase about 5% each year so generating your own electricity will become more important in the future.

The cost of PV panels have become very affordable; less than $1 per watt. A local installer/dealer can quote installation and other equipment costs. The federal tax credit of 30% still applies through 2016.

If you have the funding or can add the cost to a mortgage, it would be smart to buy the entire PV system and reap the full benefits long term. But I elected to lease my system, paying zero upfront and getting modest benefits. These benefits are obtained only if the solar cost per kWh is less than the grid costs per kWh. The lease requires a monthly payment based on a guaranteed annual kWh production. Because the sun energy is so variable, you cannot know your actual solar cost/kWh until a full year has passed. In my case, I received more than the guaranteed solar production and the actual solar cost was $.05/kWh less than the grid, saving me about $350 for the year. The full system cost

A home on Long Island, New York, with a 2.7 kW array of grid-connected PV modules. PHOTO: EVERGREEN SOLAR

would have been $27,000. I did not even pay for the building permit or the electrical inspection costs. Start your search for installers at *FindSolar.com.*

All of this tracks perfectly with a new trend (actually, an old concept with new clothing) where the emphasis is on local energy solutions. As reported in *Scientific American,* it makes more financial sense to generate new power closer to home. Wind turbines and rooftop solar panels could provide 81% of New York's power, according to the Institute for Local Self-Reliance *(www.ilsr.org).*

Local energy is not a romantic notion. Plans for the world's largest wind farm in Texas had the plug pulled because the transmission lines were too expensive. Instead, a series of smaller installations closer to major cities is planned. More and more companies are installing "micropower" plants to power a building or campus. Germany stopped subsidizing the fossil-fuel industry and enacted policies to force the utilities to buy power from local people at a decent price. They now have 1.4 million photovoltaic installations, more than any country on earth. ✺

Geothermal for Off-Gridders?

Not everyone wants or can be connected to the grid. In those off-grid cases, they would need a few more components and a battery bank for storage and backup. You should be aware, however, that GHPs use more electricity than most off-grid solar/ wind systems are designed to provide, particularly in winter when the sun's output is typically 40–50% less than in summer. If you are contemplating installing a GHP in an off-grid system, first discuss it thoroughly with a seasoned solar installer. You may just discover that the cost of solar panels and lead-acid batteries is just too high to be practical.

Ukiah, California

Doug Pratt's home in Ukiah, California, is a perfect example of marrying GHP technology with solar technology to obtain real synergy. His GHP provides four units of heat energy for every unit of electric energy to run it. "Almost like stealing," as Doug puts it. And his solar PV installation then pays for that one unit (and more) of electrical energy cost. His solar electric system delivers about 20 kWh per day when the sun is shining, and over time practically wipes out his total electric bill by running his meter backwards. He has an all-electric, passive solar home, which means that its energy-efficient design takes full advantage of the sun's energy, reducing the need for grid power. His annual electric bill: $110. I wish I could say that.

His active solar installation has 48 peel-and-stick solar panels located on the roof of his workshop; sealed batteries store the energy and two 3,600-watt inverters convert the DC solar power to usable AC power. Peel-and-stick modules are made of amorphous silicon that require about 50% more surface installation

Earth Smart Home

area to get the same amount of wattage as crystalline silicon, but can be built without glass covers for an unbreakable module.

His one-story home includes 1,450 square feet (42 sq m) with a single-zone, 2.4 ton GHP (WaterFurnace brand) connected to vertical closed loops. It has two 250-foot (76 m) deep boreholes to provide heat and cooling. The best thing about his setup? "I can make the fuel," he said. Well put, I say. His total GHP cost was $20,000 of which $12,000 was for the backyard hole drilling. That is $24 per foot, about standard.

Somehow, I cannot feel sorry for him when his biggest disappointment in the GHP is that it is so quiet that he can't tell if it is running without sticking his head in the garage. In the garage? I keep forgetting that California homes do not seem to have basements.

The garage roof holds the peel-&-stick PV panels; the inverters, charge controllers and batteries are installed inside. PHOTOS: DOUG PRATT

Chapter 12

Determining Size and Cost of a GHP System

You can get an instant price on a refrigerator, but coming up with a price for a GHP system requires an experienced and qualified contractor because there are so many variables.

Contractors / designers use Manual J computer software to determine the proper size of the GHP unit, usually expressed in tons. *(A ton is equal to 12,000 Btu of heat energy per hour. A Btu is the quantity of heat needed to raise the temperature of one pound of water by one degree Fahrenheit.)*. The GHP unit size is determined by the design heat loss for that specific home or building, and it must be large enough to replace the maximum heat lost by the home during the coldest weather. Most modest-sized homes require 3 to 5 tons. (See comments by Ed Lohrenz at the end of chapter 15 about how optimizing building efficiency can affect these calculations.)

Once the GHP size (in tons) is determined, the ground loop can be designed—again using computer software—to transfer the proper amount of heat energy from the ground to the refrigerant loop. That ground loop design is impacted by the type of soil or rock size at the home site. The final cost segment comes from choosing the method of getting the heat into the home, such as a forced-air or hot-water system.

Cost Segments

Actual costs and installation time frame will be determined the home's design and construction as well as the physical location and ground-soil conditions. A proposal will include costs for:

▸ **Hardware & Labor**: GHP equipment, transportation, desuperheater, thermostats, hot water heater, installation, etc.

▸ **The Ground Loop**: *Open-loop well-source* (piping from well to house and to GHP and return; usually excludes well costs); *Closed-loop lake source* (distance to the water is the variable factor); *Closed-loop horizontal trenching*; or *Closed-loop vertical installation*.

▸ **Heat/Cooling Delivery System**: Any system—oil, gas or GHP—can deliver heat into the home whether it is hot-water heat or hot air, but ducting air is the most popular way to get both heating and cooling.

Existing Ductwork?

If you add a shiny new GHP to an older home that already has ductwork installed, you could save a lot of money and work. Or will you? Older ductwork for forced air delivery systems was often designed to deliver air at high velocity. GHP metal ducts, on the other hand, are insulated and designed for a slower speed air delivery. This setup takes less electricity to run the blower, but it also means that larger ductwork is needed so that more air can be delivered. You should have your older ductwork inspected by your installer to determine if its size is usable and if it should be insulated.

Variables Affecting the Size of a GHP System

To determine the size of a geothermal heat pump system (in tons) as well as the size of the loop field, the heating/cooling loads (heat loss/gain) of that specific residence must be calculated. To do this, a great deal of information about the home and its location must be plugged into the Manual J software:

- Location, elevation, latitude
- Average wind speeds
- Indoor temperature settings for owner's comfort
- Local outdoor temperature/humidity ranges
- Which side of house faces north
- Window/door placement, size, material used, double- or triple-glazed, low-e value, tinted/reflective glass, etc.
- New construction vs retrofit
- House plans for size, zones, stories, insulation, tightness, ceiling heights, energy recovery systems, etc.
- Basement—finished or not; heated and/or cooled
- Garage—heated?
- Number of fireplaces
- Heat distribution system (forced air, radiant heat)
- Soil type and moisture content (for horizontal loops)
- Water-well log data, if nearby (for vertical bore site)
- Site plan: gardens, walls, poles, outbuildings, landscaping, available ground area, etc.

When receiving those first bids from the contractors, the odds are fairly high that the real numbers will give "sticker shock," but consider the following facts: a new home will require some type of heating and cooling system anyway. The real question then is: What is the *added* cost of the GHP? Keep reading for more information about calculating payback time.

Taking Advantage of Rebates & Tax Credits

The EPA has estimated that the United States alone could reduce its dependence on foreign oil by 21.5 million barrels of crude oil each year for *every* one million homes using geoexchange (as they call it) and reduce greenhouse gases by eliminating 5.8 million metric tons of CO_2 emissions annually.

In April 2010, the EPA announced new, more rigorous guidelines for earning the Energy Star label *(www.energystar.gov)*. This includes higher efficiency standards for geothermal heat pumps and included home-based solar and wind energy installations. As a part of this, the EPA also issued tax credit guidelines to help the homeowner. In order to obtain tax credits for geothermal installations, the homeowner must prove that these standards have been met and installed. **These federal tax credits will expire on December 31, 2016**.

A listing of state subsidies and rebates can be found on Geothermal Heat Pump Consortium's website: *www.geoexchange. org*. Another excellent website is the Database of Incentives for Renewables and Efficiency: *www.dsireusa.org*. ✪

....

Homeowner Insurance Savings? I had this naive idea that homeowner's insurance companies should provide a discount or credit for GHP systems, based on the lower risk of fire or carbon monoxide emissions. However, when I checked with my insurance company and several underwriters, the answer was fairly uniform. First, they were not certain what a geothermal system was. Second, they stated that insurance companies do not give discounts for heating systems.

....

Chapter 13

Payback: The Real Bottom Line

T he decision to install a GHP in a home, to expend or bor-
row scarce funds, will largely depend on whether or not
homeowners believes they will get a full return on their investment. Let's first define what we are talking about. ROI (return
on investment) is a banker's term that can dull the senses. I
like "payback" better. Payback is a term of time (years, usually)
when you finally get your money back. From that time on, you
are ahead of the game every year. You are spending money now
to save money later.

The usual calculation for payback compares the *added*
costs of a GHP over and above the cost for an oil, gas or electric
heating/cooling system, also taking into account the annual
savings that comes from eliminating the cost of fuel needed to
run a furnace and an air conditioner.

To my simple mind, payback is very clear. You are going to
pay for and install *some* type of heating system. If a GHP costs
more, but offers greater savings over time, you just need to know
when you'll break even for the extra cost.

GHP manufacturers provide a general statement: "We often
see a three- to five-year payback of the additional costs." Some

companies provide a handy savings calculator where you can plug in your data to see the savings. That may be enough for many, but you cannot see *how* they compute results or what factors they include. There are several ways to calculate your return on investment.

Model 1 – Payback on Added Costs

You spend $28,000 for a GHP system; $8,000 more than the $20,000 cost of a natural gas boiler and full AC. You will save $2,500 every year by not buying gas, but it takes $200 of electricity per year to run the GHP. So what is the payback?

> *$2,500 – $200= $2,300 energy savings per year*
> *$8,000 additional cost for GHP ÷ $2,300= 3.5 year payback*

With Subsidy: If you factor in the current 30% federal tax credit, it will cost less than a standard HVAC system:

> *$28,000 GHP cost x .30 = $8,400 tax credit*
> *$28,000 – $8,400 = $19,600 actual cost*

The payback? Zero years! When compared to a $20,000 natural gas system, the payback is immediate. The GHP costs no more than the natural gas system. In five years you will have saved another $11,500 in fuel savings and you did not even have to install a chimney! And we are ignoring the fact that any furnace/burner also requires a bit of electrical power as well as annual maintenance costs, and they have shorter life spans.

With Financing: Take the same example as above, but now also consider the cost of financing the added GHP costs. By folding the additional $8,000 into your home mortgage, your monthly payments increase slightly to cover the added principal and interest; maybe $500 per year.

> *$2,500 annual fuel savings*
> – *$200 added electricity*
> – *$500 extra mortgage costs*
> = *$1,800 savings per year*

The GHP incurs no additional cost because your total net savings starting that very first year is $1,800. And your payback now? Zero years! You have immediate payback even without applying the subsidy. And in ten years you'll have saved $18,000.

Model 2 – Payback on Home Retrofit

You are replacing your home's obsolescent oil burner with a GHP. You need both air delivery (for air conditioning) and hot water delivery for your existing radiant floor heat. You paid $4,000 the previous year for heating fuel. A new high-performance boiler and separate AC would cost $16,000 when installed. Final GHP costs are $34,000 because of the vertical boreholes for a closed-loop system. You receive a tax credit the following April of $10,200. The cost to operate the GHP is $300 per year. What is the payback?

> *$34,000 – $10,200 tax credit = $23,800 GHP cost*
> *$23,800 – $16,000 = $7,800 added cost of GHP*
> *$4,000 – $300 = $3,700 savings in fuel costs*
> *$7,800 ÷ $3,700 yearly operating costs savings = 2.1 years*

Without the federal tax credit, the payback is 4.9 years, but in 20 years you'll have saved over $74,000 if oil prices do not go up, and that is highly unlikely.

Model 3 – Payback on Total GHP Costs

Assume your total GHP system (including installation) is $30,000 and your net fuel savings per year are $1,900. Your federal tax credit will be $9,000 (30% of $30,000). A standard oil furnace with full AC would cost $22,000. What is the payback on the entire investment, not just the added cost?

> *$30,000 – $9,000 = $21,000 system cost*
> *$21,000 ÷ $1,900 = 11-year payback*

This is the most conservative approach imaginable. It means that the savings must cover the entire GHP cost! If you applied this payback concept to a standard $22,000 HVAC system, the payback would be an infinite number of years—there are no savings. But for a GHP, in 20 years the savings will have paid for the cost of the GHP plus $17,000 more.

> *$1,900 x 20 = $38,000 in fuel savings*
> *$38,000 – $21,000 system cost = $17,000 in the bank*

Don't forget that in addition to the federal subsidy (which is set to expire at the end of 2016), some states and provinces, plus electric utilities and cooperatives may also provide subsidies.

When thinking long-term, you will see that a GHP saves a lot of money. The added cost will be more than repaid and there is no pollution. Is there any better way to invest money with such a high return coupled with such a low risk? Geothermal heat pumps are *the* answer. ☼

Geothermal Savings Calculators

How much money can you save by installing a geothermal heat pump? Many GHP manufacturers' websites provide savings calculators. Just answer a few questions, such as:

▶ Nearest city/state
▶ New or existing home
▶ Square footage to be heated/cooled
▶ Insulation/air leakage: poor, average, good or excellent
▶ HVAC equipment age: 0–3 years, 3–15 years, 15+ years
▶ HVAC equipment efficiency: standard or high efficiency
▶ Heating type: forced air, radiant floor
▶ Home heating fuel: natural gas, electric heat pump, electric resistance, propane, fuel oil
▶ Water-heating fuel: natural gas, electric, propane, fuel oil
▶ Energy efficiency improvements, such as insulation upgrades, Energy Star appliances, compact fluorescent lighting
▶ Ground area available for GHP loop: horizontal, vertical

Feedback includes energy costs savings by switching to geothermal, where your energy dollars are spent (for heating, cooling, hot water, appliances, lighting), your carbon footprint and more. Also provided by some websites: Design Data for these calculations (outdoor design temp, heating degree days, earth temp, current building load, hot water usage) and Utility Rates for summer and winter.

If just one in 12 California homes installed a geothermal system, the energy saved would equal the output of nine new power plants.

Chapter 14

The Heat-Pump Cost Dilemma

When I give talks about geothermal heat pumps, I list a dozen advantages and one disadvantage, and for many homeowners, it is a big one: what will it cost?

If the homeowner can fold all equipment and installation costs into a mortgage, the payments and the benefits are both in the future and in sync, making it affordable from the start. But in a retrofit situation, a mortgage or a home equity loan is probably not available. Any federal tax credits would be realized on the following year's tax return, but the GHP costs would still need to be paid when the job is done. The fuel savings, which add up by the year, delivers benefits in the future. Therein lies the dilemma.

While US inflation has been extraordinarily low, the cost of heat pump installation has risen. Here is a specific, real-life 2014 example. Sam and Mary, both working professionals with a 3,100 square-foot home in New York State, wanted to replace an oil burner with a heat pump. They did everything right. They read my book, *The Smart Guide to Geothermal* (obviously a smart move); they corrected some problems revealed by an energy assessment; and they asked for advice from an experienced, well-qualified heat pump contractor.

The table below summarizes their contractor's bid of a complete heat pump heating and cooling installation. It shows that the 4-ton heat pump itself now costs $9,850, a 30% increase in the last 6 years (as compared to my 4-ton GHP system). Labor costs are now $75/hour for an assistant and $110/ hour for a master technician in New York. These are fair rates; you absolutely need a qualified, experienced installer. Plumbers are charging $150/hour in New York. But still, this data shows $14,000 for labor alone.

Simplified Example: 2014 New York Bid Proposal for a 4-Ton Heat Pump Installation

Hydron 410A 2-stage 4-ton GHP	$9850	Zoning system, 4 zones, thermostats	$3000
Desuperheater	incl.	Humidifier	$475
10kw Backup Heater	$375	Manifold, Piping	$500
Freight	$250	85 gal. Electric Heater	$1500
Purge Tower	$1100	Antifreeze	$50
PEX Piping, Insulation	$450	Ductwork	$5000

Material Subtotal	$22,750
Labor (tech @ $75/hr, master @ $110/hr)	$14,000
TOTAL	**$36,750**
plus 2 horizontal ground loops, estimate	*$5,000*

As you see, the total bid is $36,750 and does not cover the ground loop which may be underestimated at $5,000. The total cost of over $40,000 was too much for this family and they are not alone.

Greenbuildingadvisor.com published article on this subject in 2013. They quote Jay Egg, coauthor of *Geothermal HVAC*, who says that based on his 20 years of experience, a home system now has an average cost of $42,000. They quote comments from Maine installers whose average installed residential GHP prices run from $30,000 to $42,000. The article reasoned that loop costs have stabilized, but equipment costs are going up.

Not everyone agrees with this. Paul Auerbach of Total Green Geothermal (specializing in DX installations) tells me that while equipment costs are higher, improved DX ground-loop technology has permitted them to keep prices fairly steady.

Many homeowners are seriously considering a heat pump and contractors are expending a lot of energy in preparing detailed bids, but when that bid comes in, too often it is a "No Buy" decision.

I have come to understand that GHPs are not for everyone, but there is a huge unfilled potential: thousands of homeowners want such a system, but cannot afford it.

A Solution: Geothermal Leasing

One business model to follow is the successful solar leasing concept that is spreading widely and to which I happily subscribe. According to Forbes.com, geothermal heat pumps have the economic potential to deliver the same value: hassle-free installations and reliable savings. But if the geothermal industry is going to actually achieve this potential, it must follow three basic concepts learned from solar leasing:

One: A single company must be responsible for the entire process, from engineering to installation, maintenance, performance verification, and financing.

Two: Reliable monitoring of energy production and of

problems must exist. My solar production data is sent through my internet connection so that the company and I can monitor it. I actually received a call from the California office advising me that my inverter had a problem and they wanted to send a technician the next day. I had never noticed the problem.

Complete monitoring of heat pumps is currently available. Ground Energy Support in Dover, New Hampshire, is providing GX Tracker Monitoring systems ($395) with real-time performance information. This includes a summary of energy extracted (or rejected, for cooling) from the ground, data graphs on temperatures, flow rates and BTUs produced, and performance data.

Three: A lease-costing approach needs to be based on energy production coupled with a guaranteed annual energy production so homeowners can see the benefits from the first day.

In the solar-lease model, the annual guaranteed kWhs are the basis for the monthly lease payment. If the annual production falls below this (even if it is out of the company's control), they pay the customer. If the output exceeds the guarantee, the cost per kWh goes down and the annual savings go up.

In a similar way, heat-pump leasing program costing could be based not on efficiency, but on the BTUs produced per year.

Two other benefits to the customer make this approach attractive. In the short term, this concept provides an incentive for ongoing company diagnosis and rapid response to problems. In the long term, the provider has an incentive to prepare for the future replacement of parts and subsystems, and even upgrades.

As Forbes puts it: "We already have the corporate and financial structures to bring the solar-lease model to GHPs. In fact, it takes little stretch of the imagination to see a solar lease company acquiring or partnering with GHP installers and offering a geothermal lease along with the solar lease to its customers."

Another solution is for the power utilities, local municipalities, or developers to own the loops or even the entire system, lease it to the homeowner, and recover the capital investment with an on-bill repayment over the life of the system. For example, the utility would install, maintain, and own the GHP loop-piping network while the customer would own and maintain the heat pump. Or they could lease the entire system. The utility would charge customers either a monthly fee or a usage charge.

The inset below describes a Minnesota electric cooperative that is actively providing a leased GHP service, treating it as just another energy utility.

Novel Approach to Geothermal by Electric Cooperative

The Lake Region Electric Cooperative (LREC) with 25,000 members in Minnesota takes the word "Cooperative" seriously. They have initiated a unique business plan (called the EarthWISE Geothermal Program) that eliminates the initial costs of geothermal underground loop systems for homeowners who install geothermal heat pumps. For each home under this plan, LREC will custom design (using LoopLink software), install, and pay for the underground loop system. They would then charge a flat monthly fee (loop tariff) for a 35-year lease of the loop system. Alternatively, the homeowner can buy the loop at any time.

LREC does their own installation (up to the house), using horizontal directional drilling that requires less ground disruption.

By lowering the cost of installation, using local and federal incentives, and by reducing annual heating costs, a GHP system becomes very affordable. A solid proponent of geothermal ground source energy, LREC has also installed their own GHP system in their Operations Center.

PACE Program

PACE (Property Assessed Clean Energy) is a national organization that provides models, guidelines and procedures to states, counties and municipalities on how to extend the use of land-secured finance districts to fund energy improvements on private property. Heat pumps and solar panels can be included. Their motto is "Funding Energy Efficiency." The DOE provides guidelines for setting up pilot programs. The PACE program pays for 100% of a project and is repaid over 20 years with an assessment added to homeowner's property tax bill.

State legislation is required to enable PACE programs. Thirty-two states have passed the legislation and 426 programs are currently underway. For example, the CaliforniaFIRST program is a partnership between counties and private financing companies that requires no money down for homeowners and commercial operations. *Scientific American* magazine has named the PACE program as one of the "20 World Changing Ideas." To check a specific state, go to *www.pacenow.org*.

Other Solutions Available Now

A number of interesting financing options are available right now. Here are just a few:

The Plumas-Sierra Rural Electric Coop in Portola, California, offers an interest-free, 15-year loop-loan program up to $15,000 to cover the ground-loop costs. The monthly repayment is $29.90 for a horizontal loop and $83.30 for a vertical loop. In addition, they provide a free electric water heater.

Massachusetts' residents have a Mass Save Heat Loan program that provides 0% loans for GHP systems up to $25,000 for 7 years. A $40,000 system cost would have a $12,000 federal credit, a $25,000 loan at 0%, paying $298/month for 7 years.

A Canadian non-profit organization in Nova Scotia (Efficiency Nova Scotia) has partnered with a bank to offer a 0% loan for GHP ground-loop installations up to $15,000 for 5 years.

As the US federal subsidy program ends in 2016 with little chances of renewal, I believe that programs such as those described will become even more important for the adoption of this proven technology. ✪

Specialized high-speed, lightweight drilling rigs are often used to punch the vertical boreholes. Here you see a loop pipe being inserted into a borehole. PHOTOS: DOUG PRATT

Chapter 15

Finding and Selecting a Contractor

A heat pump installation requires a higher level of knowledge and experience than an oil or gas burner installation. The GHP may be simple in concept, but complex in practice. You do not plug a GHP into a wall socket. It has feedback loops, safety controls, fault sensors, shutoff circuits, diagnostic LEDs, and a large microprocessor control board. (Amber LED S-12 is blinking—what does that mean?) The contractor must consider the entire system, from the heat source to the final comfort of the occupants. While GHP contractors may also install heating, ventilation and air conditioning (HVAC), not all HVAC contractors are experienced with GHP systems; they are fewer in number and they are getting busier.

The goal of any homeowner, general contractor, or architect is to find an experienced local geothermal contractor and a high quality geothermal manufacturing company with the latest technology. There are two ways to do this: search locally for a contractor/installer and use his judgment and experience in finding a manufacturer, or select a manufacturer and use their website to find a qualified, accredited contractor.

Search for a GHP Manufacturer

If you are searching for a manufacturer, review the appendix list of US and Canadian manufacturers, find out how long they've been in business, and then make sure they:

▸ **Have qualified the hardware** to US EPA (Environmental Protection Agency) Energy Star standards, or ISO 13256 (International Organization for Standards) certification, or Canadian CSA C13256 (Canadian Standards Association) standards. See chapter 21 for more information on standards.

▸ **Provide the latest technology** in high-efficiency compressors (two-stage, digital scroll compressors, etc.).

▸ **Have good efficiency ratings.** Performance is measured in terms of efficiency (COP for heating and EER for cooling). Definitions of COP and EER are described in chapter 20. The US Energy Star initiative publishes each year the most efficient heat pump products *(www.energystar.gov).* COP ratings for heating should be:

    Closed-loop with Hot Air ≥ 3.6

    Closed-loop with Hot Water ≥ 3.1

    Open Loop with Hot Air ≥ 4.1

    Open Loop with Hot Water ≥ 3.5

    DX ≥ 3.6

▸ **Offer a solid, comprehensive warranty.** Warranties make us feel good, and if the system fails (Murphy's Law says it will always happen in a severe snowstorm), they become very important. But this should not be the sole criterion used. For one thing, if the system survives the first few years, it will probably keep working for 20–25 years. Service and GHP efficiency are also very important. I would be concerned, however, if no warranty is offered. The EPA Energy Star warranty requirements are a minimum of 5 years for major components (compressor, heat exchangers, expansion and reversing valves, and air coils); other parts, 2 year minimum.

▸ **List the names of qualified contractors in your area.**

Search Locally for a Contractor/Installer

If searching online or by telephone directory for a local contractor/installer, consider that the International Ground Source Heat Pump Association (IGSHPA) has a list of accredited contractors, listed by state or Canadian province: *www.igshpa.okstate.edu*. When you find a contractor, inquire about the following:

▸ Have they installed a GHP in their own home?

▸ Are they an accredited installer/contractor?

▸ How many GHP systems have they installed, both open loop and closed loop?

▸ Do they use load-calculation software to determine the correct size of unit for your home and the loop field needed? *(This calculation is very important—insist on it. The time has passed when back of the envelope estimates are good enough.)*

▸ Do they have trenching or drilling equipment or do they work with a capable subcontractor with years of geothermal experience?

▸ Will they provide a written statement of cost, delivery and warranty?

▸ What warranty do they offer on the equipment *and* the installation? *Get this in writing.*

▸ Which manufacturer's equipment are they using? How long have they used this equipment? What do they like about that manufacturer?

A three-day accreditation course is a necessary but insufficient measure of installation expertise. Trainer Michael Hunt has found that he has to work with the majority of class participants for a year or two after the class to get them on the right path. There is no substitute for experience and mentoring. A contractor in your area is going to be better at sizing and designing the complete system than one many miles away without geological knowledge of the area. If your installation requires drilling boreholes, the contractor should either have his own drilling rig or subcontract for it.

Other Questions to Ask

Ask about desuperheaters and about tankless, "on demand" water heaters. What about Slinky loops? Ask about the payback, efficiency, operating costs and system life span.

Get references. Call them. This cannot be overstated.

The contractor should run software programs to determine the size of the unit needed and then the amount of ground piping required to obtain the ground energy. These calculations involve the house size, tightness, insulation, windows, and number of zones to determine the design heat loss. The analysis should also include the degree-days, the winter and summer indoor design (preferred) temperatures, seasonal outside average temperatures, geology of the area, type of ground, size of land area, and even the elevation. What works in Winnipeg will be different from Wyoming.

In my case, it worked out to about one ton of capacity for each 600 square feet (56 sq m) of house. Our 2,400 square-foot home (223 sq m) required a 4-ton unit and that has proven perfectly adequate for our land and weather. My elevation is only 249 feet (76 m) above sea level, although I can see the 4,000-foot Catskill Mountains across the Hudson River. Move my home anywhere else and the unit size may be higher or lower.

Some contractors discourage the use of open-loop systems because the water is less controllable and local codes must be followed. Listen to them, but consider the lower installation cost benefits. Other contractors will look for open-loop possibilities as a first choice, since that is the most efficient and least costly to install. They will also determine if there is real estate available for a horizontal ground loop, and finally vertical holes.

It is very important to not oversize the unit by installing one too large for the load in your home. If the GHP ton-rating is higher than needed, it could start a "short cycle" where frequent on/off cycling of the cooling system would occur as it reaches the thermostat's setting too quickly. This can shorten the life of the unit. Oversizing the ground loop, on the other hand, can be an advantage especially if there are plans to increase the house size in the future.

Hire a contractor familiar with the local topography. Shallow ground temperatures are relatively constant on this continent, therefore geothermal heat pumps can be used effectively almost anywhere. But determining the best type of ground loop suitable for the composition and properties of your soil and rock require experience. Soil with good heat transfer properties (i.e. soil with moisture) requires less piping to gather the heat energy needed. If extensive rock ledge is present, vertical ground loops may be required. Ground or surface water availability may be suitable for an open- or a closed-loop system. ✪

ADVICE FROM THE EXPERTS

Optimize Building Efficiency. Ed Lohrenz, founder of GEOptimize in Winnipeg, Canada, and instructor and developer of the IGSHPA Certified GeoExchange Design Course, stresses the importance of heat loss/gain calculations and their impact on the size and cost of the heat pump system.

He recommends to first do the calculation to determine the size of the heat pump and ground loop. Then work with the owner, builder or architect to discuss small changes in the building that could make large changes in the heat pump system. Factor these changes into the calculations and see what happens. He has found that it is possible to reduce the ground loop size by 10 to 20% and reduce the heat pump size from 4-ton to 3-ton units. Changes to a building include:

▸ Adding insulation to the attic, basement, and crawl spaces.
▸ Replacing old windows with 2- or 3-pane argon-filled units or adding reflective window films.
▸ Reducing infiltration by sealing the area where the floor joists connect to the concrete foundation.
▸ Increasing the roof overhang on south-facing glass.
▸ Installing a heat-recovery ventilation system.

His experience shows that this approach can reduce energy consumption and lower GHP installation costs. It may even permit the installation of a horizontal ground loop instead of a more expensive vertical loop.

Understand How Loops Affect Ground Temperature. Lawrence Muhammad of Geo NetZero in Michigan cautions installers to be aware of the danger of placing vertical boreholes too close together. He has seen installations where they have been as close as 10–12 feet. This works at first but over time, the ground temperature will rise several degrees, changing the water temperature in the ground loop and the overall efficiency of the system.

Compensate for the Climate. Judith Karpova, an IGSHPA-certified installer of Green Pathways Consulting in New York, raises a point about open-loop systems where a well is the source of the entering water temperature (EWT) and the discharge source. When we experience a warmer climate for longer periods of time, the aquifer does not have time to recover and the EWT is unable provide cooling. Likewise, during prolonged severe winters, the aquifer EWT is too low to provide heat, shutting down the system. The solution is to provide a temporary alternative for diverting the water until the aquifer can recover.

Do Good Proposals. A thorough proposal is a good indicator of the contractor's work. A proposal should leave nothing to chance or misinterpretation: it should spell out the exact equipment, subcontractor work, costs, schedules, statement of exactly what is to be done, warranties for the hardware and ground loop, drawings, timing, payback over the years, permits, descriptions, and pictures of the equipment. It should include sign-offs for the homeowner for permission to remove shrubbery, locations of trenching, wells, etc. *Geothermal HVAC—Green Heating and Cooling* by Jay Egg and Brian Howard provides an excellent model of superior proposals and why they are important.

Go the Extra Mile. Egg Geothermal in Florida has a policy to leave a yard in better shape than they found it. Anything that was damaged is replaced, down to new sod and mulch in flower beds. This is in contrast to some installations where homeowners have complained their entire yards were destroyed. Horizontal boring has been another blessing in this regard.

Part II

Unraveling the Science and Technology of Ground-Source Heat Pumps

*How can you possibly start with a
low temperature, maybe 50 degrees F,
and raise it to 165 degrees F, without burning
something or using magic?*

That question hits at the very heart of a heat pump, and accomplishing this neat trick is what separates heat pumps from all other forms of heating and cooling.

How does it work? In the following three chapters we'll learn how a heat pump takes clever advantage of a number of the physical laws of nature to transfer heat energy and to raise temperatures.

The most critical action takes place in the refrigerant loop, shown in a water-to-air system in heating mode in Figure 16-1. The "magic" is best explained in three steps:

1) Chapter 16 gives a quick overview of a single cycle of the Refrigerant Loop, showing step-by-step **what** is happening at each critical point where change is taking place.

2) Chapter 17 gets into the **why** and **how** of heat pumps. You cannot really visualize how heat energy is transferred and temperatures are increased in a heat pump until you understand the science.

3) Chapter 18 puts it all together by showing temperature and pressure changes inside that loop so we can see the transfer of heat energy in a slow-motion single cycle.

Chapter 16

A Look Inside: How a GHP Delivers Heat into the Home

I know from my own experience that a GHP is clearly able to heat my home very comfortably regardless of the outside temperature. To do this in a water-to-air system, 165°F (74°C) is usually required at the final heat exchanger so that cool air can be heated, travel through the ductwork and enter rooms at about 100–105°F (38–41°C). In a water-to-water system, only 120°F (49°C) is needed to deliver radiant or baseboard heat.

I use 50°F (10°C) as an average ground temperature only as an example to show the basic theory of why things work. Ground temperatures vary greatly depending on geography, climate conditions, and many other factors that impact GHP design and installation.

There are two basic ways to increase the temperature of a substance in a closed system. First, when heat energy is introduced into a substance, its temperature will rise. If more than one source of heat energy is introduced to the substance, each additional source will raise the temperature even further.

The second method is to increase the pressure. Liquids cannot be pressurized, but gases can be. Refrigerants are used because they can easily change state from a liquid to a gas and back to a liquid (a phase change). Increasing the pressure will increase the temperature of a gaseous substance proportionately.

When we pry open the door to look inside a GHP, we'll see that both of these methods are at work. It all happens in a number of incremental steps to achieve that final working temperature.

1) Earth Heat Energy

The first transfer of energy takes place when the sun's energy is absorbed by the earth, combined with the gradient heat radiating from the core (albeit 4,000 miles away), keeping the ground at a relatively constant level. Even though it feels cold to your hand, any substance at 50°F (10°C) actually contains a great deal of heat energy that can be transferred.

The field of thermodynamics studies the behavior of energy flow and as a result, physical laws have been written to explain it. The Third Law of Thermodynamics helps us to understand that any body even at 50°F contains heat energy: it is over 500 degrees (F) above absolute zero, a theoretical condition in which no heat energy exists in a body. In other words, there is more than enough heat energy in the earth that can be continually captured and directed into the home.

2) Ease of Transfer

The next transfer of heat energy is from the earth to the ground loop of a geothermal system, which raises the loop's water temperature. Then, in the next increment, that heat energy is transferred to the refrigerant loop, thus increasing its temperature and changing it into a gas.

Those energy transfers are easy because of help from Mother Nature. Heat energy flows spontaneously from an area of high concentration (hot body) to an area of low concentration (colder body). The greater the difference, the greater the transfer rate. Of course, this is also what's happening when your cup of tea cools off or an ice cube melts.

When you dip your fingers into 50°F water, it feels cold because there is a 48.6-degree difference in temperature between the water and your body. Heat energy is moving from your fingers into the water, raising the temperature of the water, if ever so slightly.

All of this is embodied in the Second Law of Thermodynamics.

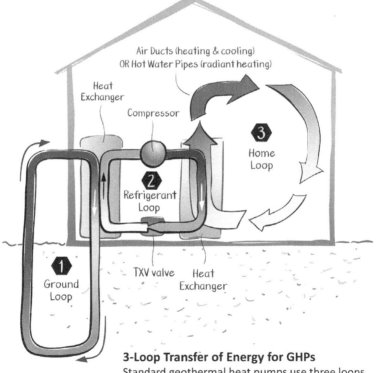

3-Loop Transfer of Energy for GHPs
Standard geothermal heat pumps use three loops to transfer heat from the earth to the home.

Figure 16-1

The temperature *difference* is the key: as long as there is a temperature differential, heat energy will spontaneously flow to the colder object, raising its temperature.

A ground loop containing a flow of water at, say, 33°F (0.6°C) will continuously pick up heat energy from the earth—that is at 50°F (10°C)—and the loop's temperature will increase. This heat energy will be continually passed into the refrigerant loop through a heat exchanger, increasing the temperature of the refrigerant in that second loop.

3) Compression

Next, the refrigerant gas is sucked into a device that increases the pressure (that's the compressor, of course), and the temperature of the refrigerant increases even further, proportionate to the pressure increase.

If you've ever used a household pressure cooker, you know that after you add food and water, close it up and turn on the heat, the water starts to boil and turns into a gas. The pressure increases inside the closed cooker and the cooking time is reduced.

This effect is based on what is called the Ideal Gas Law. It states that in any closed, fixed-volume system, the pressure (P) is proportional to the temperature (T). In other words, if P

What is a Refrigerant?

Refrigerant fluids are used in heat pumps to absorb heat energy at low temperatures and release heat energy at higher temperatures. The HVAC industry has given trade names to EPA-certified refrigerants in order to identify different chemical makeups. Since R22 (Freon) has been linked to ozone depletion, it is now banned from manufacture. R410A (Puron) has become the standard replacement in the industry.

increases (or decreases), then T must increase (or decrease). If P doubles, then T doubles.

The compressor in a GHP applies that pressure/temperature law by compressing the gaseous refrigerant, increasing the pressure and increasing the temperature. This becomes another step that further increases the temperature of the refrigerant.

4) A Second Source of Heat Energy

And finally, another source of heat energy is introduced into the refrigerant. The electrical energy that drives the compressor is converted to mechanical energy and then into heat energy. That heat energy also increases the temperature of the refrigerant. Now we have two sources of heat energy, both of which are increasing the temperature of the refrigerant. The geothermal heat pump has now achieved super efficiency and we have high enough temperatures—165°F (74°C)—to heat the home.

5) The Final Step

Back to that refrigerant: The refrigerant is in a closed loop, but we have not yet finished that loop. The refrigerant must finally be allowed to reduce its pressure (and therefore the temperature) by expansion. That expansion is the job of the thermostatic expansion valve (TXV). It performs the opposite function of the compressor by allowing the high pressure refrigerant to expand, thus lowering the pressure and lowering the temperature considerably, sometimes below freezing.

This is exactly what is needed because the lower the temperature, the greater the temperature differential in the ground loop and the more efficient the heat energy transfer becomes. ✪

Chapter 17

Simple Heat Science

Before we delve into the mechanical workings of the refriger-ant loop of a geothermal heat pump, let's take a look at the physical laws working quietly behind the scenes. All of these are critical to the movement of heat energy from the ground into your home. These include:

▶ Sensible and Latent heat transfers
▶ Laws of Thermodynamics
▶ Ideal Gas Law

First, to review:
1) Energy is defined as the ability to do work and comes in many forms: electrical, mechanical, heat, nuclear, light, etc.
2) Heat is the total thermal energy in a substance and is mea-sured in calories, Joules or Btu. It can be transferred from one substance to another. In a heat pump it is transferred by conduction and convection.
3) Temperature is the measure of the average molecular motion of a substance. It is measured in degrees Fahrenheit (F) or Celsius (C). Heat energy and temperature are related, but not the same.

Phase Change Phenomenon—Energy Storage from Latent Heat Transfer

There are three primary states or phases of matter: Solid, Liquid, Gas. When a substance changes from one state to another, we say it has undergone a phase change. These changes always occur with a latent transfer of heat energy. Heat energy either comes into the material during a change of phase, or heat energy comes out of it. Although the heat energy content of the material changes when in the transition stage, the temperature does not.

The **five phase changes** are shown below:

Phase Change	Term Used	Heat Energy Movement	Temperature Change
Solid to Liquid	Melting	Heat goes into solid as it melts	None
Liquid to Solid	Freezing	Heat leaves liquid as it freezes	None
Liquid to Gas	Vaporization, Boiling, Evaporation	Heat goes into liquid as it vaporizes	None
Gas to Liquid	Condensation	Heat leaves gas as it condenses	None
Solid to Gas	Sublimation	Heat goes into solid as it sublimates	None

Figure 17-1 Five Phase Changes

So, how can there be a change in heat energy during a state change without a change in temperature? During a change in state, heat energy is used to change the bonding of molecules instead of changing the temperature. Going from a liquid to a gas for example requires energy to break the bonds between molecules. Later, when changing from a gas to a liquid, the heat energy leaves the liquid to increase the molecular bonding.

Water is a good example to demonstrate how heat energy and temperature are related (Figure 17-4). As heat energy is introduced into a volume of water, the temperature increases until it reaches the boiling point. Then the curve flattens out. Heat energy is still being applied but the temperature stays constant. Once the water is changed into a gas, the temperature again increases. This process is completely reversible.

In a heat pump, two phase changes are applicable as latent heat transfers, as shown below.

Heat Pump Phase Changes	Term Used	Heat Movement	Temperature Change
Liquid to Gas	Vaporization, Boiling, Evaporation	Heat goes into liquid as it vaporizes	None
Gas to Liquid	Condensation	Heat leaves gas as it condenses	None

Figure 17-2 GHP Phase Changes

Latent vs Sensible Heat Transfer

When you transfer heat energy to a substance such as a liquid refrigerant, one of two things can happen:

1) As we just reviewed, the substance can change its state or phase (liquid to gas, for example) because the molecular bonds that bind the substance and determine its state are being released or strengthened. Energy is still being absorbed and stored, but it is hidden. With **latent heat** transfer, there is no change in temperature. We cannot measure it with a thermometer; we can only observe the results.

2) A rise in temperature because the molecules are vibrating, rotating or moving faster. Heat energy is being absorbed,

and temporarily stored and the temperature change can be measured—**sensible heat**.

We are accustomed to hot materials cooling off by transferring their heat energy to nearby substances. We can feel the hot water warming up the teacup. We could measure the temperatures of both. That is called sensible heat transfer.

Sensible heat can be sensed or measured with a thermometer. The addition or removal of sensible heat will always cause a change in the temperature of the substance. It will not cause a change in the physical state of the substance.

Latent heat affects the physical state of things; sensible heat affects the temperature of things. Both of these are crucial to the ability of a heat pump to transfer heat energy from the ground to the home, or from the home to the ground. But, as you will see, latent heat transfer is the dominant factor in heat pump.

The latent heat transfer plus the sensible heat transfer in

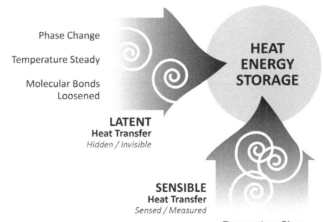

Figure 17-3
Latent/Sensible Heat Transfers

the evaporator heat exchanger, for example, produces about 75% of the energy capacity needed to heat or cool your home.

The Role of a Refrigerant

So why not use water as a refrigerant in a GHP? It is a natural refrigerant, harmless to man and nature, and readily available and disposable. Actually, it is used in some chillers, but the problem is that it requires 200 to 500 times more volume flow, plus higher pressures and higher compressor speeds. Therefore more practical refrigerant materials have been developed and are used in heat pumps, air conditioners, and refrigerators.

Exactly how does a refrigerant perform differently from water? Figure 17-5 shows the phase change of the common refrigerant R410A. These data points were taken under different conditions than inside a heat pump, but they clearly demonstrate the greater efficiency of a manufactured refrigerant. The R410A refrigerant transitions far more quickly (and

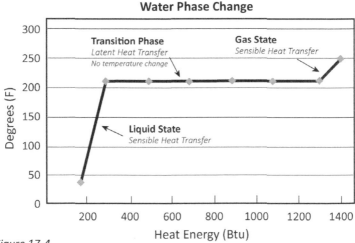

Figure 17-4
Phase Changes for Water

with less heat energy) to a gas than does water. Note that the temperature starts to increase as the transition from a liquid to a gas begins. As the heat continues to be added, a vapor phase change is underway. The temperature remains constant as the heat breaks the molecular bonds in the liquid.

This process is also reversible. The energy stored in the refrigerant is released in the condensing heat exchanger as the energy is transferred to the cool air.

The importance of the refrigerant in a heat pump (or air conditioner or refrigerator) and the ability of that refrigerant to easily change from a liquid to a gas and back cannot be overstated. The system would not work without this capability. This ability to absorb (and later release) large amounts of heat energy as a substance changes state (or phase) is a critical step in transferring heat energy "uphill" from the ground to the home.

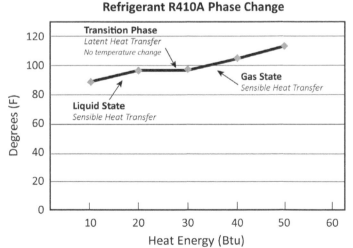

Figure 17-5
Phase Changes for R410A Refrigerant

Laws of Thermodynamics

The laws of thermodynamics describe some of the fundamental truths observed in our universe.

Ground temperatures around North America vary considerably, from 42°F in Calgary, Canada, to 78°F in Miami, Florida. Common wisdom would say this level is too low to possibly heat a home. But even at 42°F the earth contains a huge amount of heat energy that is constantly renewed by the sun as well as heat energy from the earth's core. The **Third Law of Thermodynamics** tells us that a substance at zero degrees Kelvin (-469°F / -278°C), absolute zero, contains no heat energy at all. (This is theoretical since no substance in the universe ever gets that low.) The earth at 42°F is over 500°F above absolute zero and contains a large amount of heat energy.

The **Second Law of Thermodynamics** deals basically with entropy or the measure of disorder in a system. In brief, it states that heat energy will always transfer spontaneously from a hot body to a colder body, but never the reverse. As long as there is a difference in temperature, heat energy will flow in one direction. The greater the temperature difference between two bodies (such as the ground and the water in the ground loop pipe), the greater the flow rate and the better the heat transfer. This is based on Fourier's Law, the law of heat conduction and is important in the design of the ground loop.

The **First Law of Thermodynamics** (often called the Law of Conservation of Energy) tells us that energy cannot be created or destroyed. It can be stored or transformed into other forms of energy. Storage was described above in latent heat, for example.

Heat energy is also obtained by transforming other forms of energy such as electrical or mechanical energy into heat energy.

In a heat pump, 25% of the energy capacity needed to heat or cool your home comes from a transformation of electrical energy to mechanical energy and then to heat energy.

How the Ideal Gas Law Explains the Refrigerant's Rise in Temperature

In a closed-loop system, the same water (or refrigerant) is constantly recirculated within sealed pipes/tubes. In the refrigerant loop, pressure and temperature are related almost directly: double the pressure and you double the temperature.

Here is where the heat pump compressor pays for itself: it compresses the refrigerant gas, which raises the temperature. It also keeps the refrigerant circulating so that these heat transfers are continuous.

We know from experience with a pressure cooker that liquid in a closed system becomes a gas when you add heat, while the gas pressure and the temperature both increase. This pressure/temperature relationship is described in the Ideal Gas Law:

$$\frac{Pressure}{Temperature} = Constant$$

Figure 17-6 shows the Ideal Gas Law and the straight-line relationship between pressure and temperature for water. The Ideal Gas Law, however, is for a monatomic, theoretical gas. How different would a compound refrigerant such as R410A behave? Well, not too different.

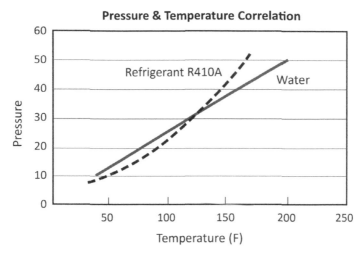

Figure 17-6
The Ideal Gas Law demonstrates a straight-line correlation for water, and a very similar relationship for the refrigerant R410A.

As you can also see, the response is not quite a straight line, but there is still a definite correlation. When the compressor does its job of increasing the pressure, the temperature increases almost directly. ✪

Chapter 18

How a GHP Creates a
Home Comfort Zone

Finally, it is time to focus on that middle refrigerant loop. We will review what is happening to the temperature, pressure, and phase changes in one full cycle of the loop in a water-to-air heat pump refrigeration cycle in the heating mode. This 3-step schematic allows us to examine the four stages of a GHP:

▸ Figure 18-1 shows the hidden substructure
▸ Figure 18-2 adds pressure and temperature changes
▸ Figure 18-3 adds sensible and latent heat transfers

You might ask why 165°F is needed at the final air heat exchanger when a home only needs the thermostat set at 70-72°F? First of all, in an air delivery system any temperature less than skin temperature will feel uncomfortable; 100-110°F is ideal. Secondly, the insulated air supply ductwork throughout the home has losses and takes away some of that heat.

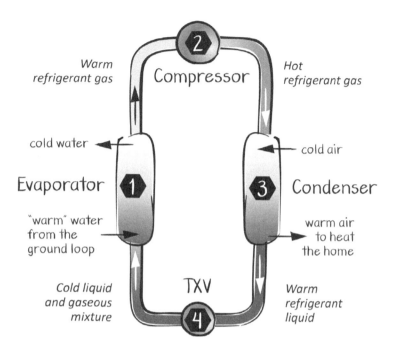

Figure 18-1

Inside the Refrigerant Loop
Hidden substructure of the middle refrigerant loop
of a water-to-air system in heating mode.

#1 – Evaporator

28°F Liquid / Gas Phase

The evaporator heat exchanger is our starting point with the
flow of a cool liquid refrigerant (with a small amount of gas) at
28°F (-2°C). Now, warmer water from the ground loop transfers
heat energy into the evaporator refrigerant. The Second Law
of Thermodynamics is at work—a spontaneous flow of energy
from a warmer substance to a cooler substance.

28°F Gas Phase

At this point, a transition (phase change) is taking place. The refrigerant liquid is changing into a gas. The temperature is not increasing; instead, a large amount of heat energy is being used to release the molecular bonds of the liquid (a latent heat transfer). In effect, this acts as an energy storage device until it is released later. In this critical step, the heat pump takes advantage of that energy stored during phase changes. The energy content

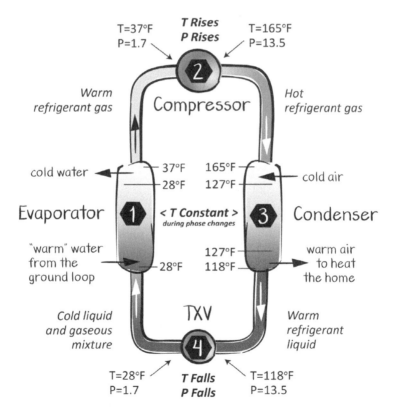

Figure 18-2

Pressure & Temperature Changes Inside the Refrigerant Loop
The refrigerant loop, with pressure and temperature added.
Source: Karl Oschsner

changes dramatically while the temperature remains constant; 70% of the total heat energy comes from this action.

Another way to look at this phenomenon is to recognize that this is following the First Law of Thermodynamics in that energy is being stored.

37°F   Gas Phase

When the phase change is complete, the heat energy from the ground is now able to increase the temperature of the gaseous refrigerant to 37°F (3°C) in a sensible heat transfer. This is a small contribution to the total heat transfer—about 5%. The refrigerant now exits the evaporator. The ground-source water has been cooled and returns to the ground for reheating.

#2 – Compressor

165°F   Gas Phase

The warm 37°F refrigerant gas is now sucked into the compressor and the Ideal Gas Law goes into effect. As the compressor increases the pressure, the temperature increases. As a result, the temperature rises to 165°F (74°C), an increase of 345%. The temperature is now ideal, but the energy level is not yet high enough.

An additional boost comes from the electrical energy that runs the compressor; it is first converted to mechanical energy to run the compressor, and then is converted into heat energy. The system now has the proper temperature and 100% of the required heat energy for the home.

Interestingly, this same phenomenon is responsible for heat pump super efficiency, or more accurately, the coefficient of performance (COP).

In summary, the compressor accomplishes a number of tasks: it provides energy to circulate the refrigerant, it raises the refrigerant temperature, and finally, it supplies added heat energy to the system while providing COPs of over 400%.

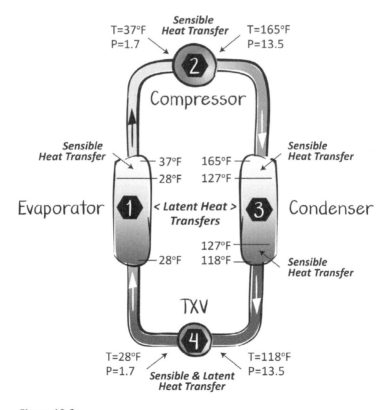

Figure 18-3

Sensible & Latent Heat Transfers Inside the Refrigerant Loop
Hidden substructure of refrigerant loop, showing the sensible and latent heat transfers.

#3 – Condenser

127°F   Gas Phase

The hot (165°F) refrigerant now enters the condensing heat exchanger and is cooled to 127°F (53°C) as the home receives that heat energy. Again, heat energy is flowing from the warmer to the cooler substance in a sensible, spontaneous heat transfer.

The refrigerant now transitions from a gas to a liquid while the temperature remains constant in a latent heat transfer. A large amount of stored energy is released as the molecular bonds are restored. The molecular bonding energy has been transformed into heat energy.

118°F   Liquid Phase

The change of phase from a gas to a liquid is completed. The temperature of the liquid refrigerant falls to 118°F (48°C) as additional sensible heat is transferred to the air. The air loop now has received a full transfer of heat energy at the proper temperature to provide room comfort. Heat energy has again flowed from the warmer to the cooler substance, and 100% of the required heat energy has been transferred into the home's air delivery system.

#4 – Expansion Valve (TXV)

28°F   Liquid/Gas Phase

The expansion valve acts like a dam with a hole in the side; high pressure on one side and low pressure on the other. The Gas Law prevails: as the pressure drops, the temperature drops and the liquid refrigerant is now cooled to 28°F. But where has the heat energy gone?

If we could look inside the TXV we would see it uses no electrical energy, but it has moving parts. A thermostatic bulb receives heat energy and activates a diaphragm that controls the size of the orifice (the hole in the dam). This is heat energy being transformed into mechanical energy.

But a larger effect is that some of the liquid refrigerant is forced to transition into a gas (called flash gas) as it exits the orifice, thus gaining heat energy, allowing the temperature to drop in what is now a mixture of liquid and a small amount of gas in a latent heat transfer.

The cycle is now ready to start over. ✪

The EPA has stated that high-efficiency geothermal systems are on average 48% more efficient than gas furnaces, 75% more efficient than oil furnaces, and 43% more efficient when in the cooling mode.

Chapter 19

Super Efficiencies of Ground-Source Heat Pumps

"*I was taught that nothing could achieve 100% efficiency. Are 500% efficiencies real or is this a marketing tool?*"

Well, actually they are both. We can demonstrate that these are real, achievable efficiencies (my own home, for example). But heat-pump super-efficiency is the single most powerful reason for buying a heat pump. That high COP level is what provides the annual savings, year after year, that separates ground-source heat pumps from every other form of heating and cooling.

The purpose of this chapter is to provide the clearest way of explaining COP and super efficiencies to those at any level of experience from homeowner to student or trainee.

The oil guzzlers in our old schoolhouse could achieve a maximum of 86% efficiency by burning (changing the state of fuel oil into heat) and this was only after a fairly expensive annual tune-up. Where did that 14% go? Most of it went up the chimney as wasted heat energy and pollutants.

The efficiency of any machine is measured by the degree to which friction and other factors reduce the actual work output.

When changing the state of a material by combustion, the output cannot equal or exceed the input or we would have perpetual motion. So the value will never reach 100%.

$$Efficiency = \frac{OUTPUT}{INPUT} < 100\%$$

It has no dimensions, just a percent, but because the input has a financial cost, it can also be expressed more realistically as:

$$Efficiency = \frac{Energy\ You\ Actually\ Get}{Energy\ You\ Pay\ For}$$

In an apparent disagreement, however, the US Department of Energy has stated that ground-based geothermal systems achieve *400 to 600% efficiency*. An efficiency of 500% means than the energy *out* is 5 times larger than the energy *in*. How is that possible?

It really amounts to a difference in definition, since the word "efficiency," in the strict sense, is used differently for a heat pump. Although it is still Output divided by Input, heat pumps (in their heating mode) are measured by a coefficient of performance (COP). These devices are moving heat and not burning an energy source to get heat, so the amount of heat they *move* can be greater than the actual input energy (electricity). Therefore, heat pumps are actually a far more efficient way of heating than simply converting fuel into heat, as in a furnace. Figure 19-1 provides a graphic look at how super efficiency is obtained in a GHP.

That illustration now provides the look inside a heat pump. It shows that energy harvested from the ground is "free" so it is only considered on the Output side and not in the Input side of the equation. The important point is the Output is the sum of the Input electrical energy *plus* the free heat energy from the ground. As described in Chapter 17, this is because the electrical energy is first converted to mechanical energy to drive the compressor, and then into heat energy. Energy cannot be created or used up, but it can be easily transformed into other forms.

Output = Input Electrical Energy + Ground Heat Energy

As a part of the COP definition, the Input electrical energy in kilowatt-hours (kWh) is converted to Btu (1 kWh = 3,412 Btu). Now, all parts are in the same units. Our graphic illustrates a simple equation in a GHP heating mode:

$$COP = \frac{OUTPUT\ (Btu\ of\ Electric\ Energy + Btu\ of\ Ground\ Energy)}{INPUT\ (Btu\ of\ Electric\ Energy)}$$

The numerator (Output) must therefore always be greater than the denominator (Input) and in practice runs from about 3.5 (or 350%) to about 5 (or 500%), or even higher.

So, we have transformed the graphic into a simple COP equation, which in turn illustrates what is actually happening inside the GHP refrigerant loop. Next, a look at efficiency standards. ✪

....

Robert C. Webber, an American inventor, is credited with building the first heat pump during the late 1940s.

....

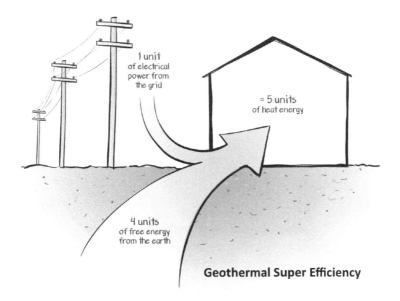

$$COP = \frac{\textbf{\textit{OUTPUT}} \textit{ (1 Input Unit + 4 Free Heat Energy Units)}}{\textbf{\textit{INPUT}} \textit{ (1 Input Unit)}} = \textbf{\textit{500\%}}$$

Figure 19-1
Heating Efficiency = Coefficient of Performance (COP)

Mechanical Advantage Analogy: The Lever

From an energy standpoint, a GHP works in a similar fashion to ancient mechanical devices such as the lever, the inclined plane, the pulley, or the wedge. These simple devices have a mechanical advantage in the same way a GHP has an energy advantage.

A lever is one of those machines with an inherent mechanical advantage that allows a person to move a much heavier object than they would normally be able to. Archimedes stated the lever's mathematical principle in the 3rd century BC. "Give me the place to stand and I shall move the earth." In Egypt, the lever was used to lift and move 100-ton obelisks.

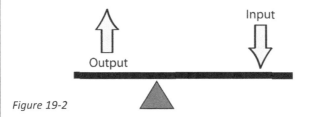

Figure 19-2

The ratio of the energy output of a mechanical system divided by the energy input is the measure of mechanical performance. Sound familiar? If the length of the lever to the right of the fulcrum is 3 feet (0.9 m) and to the left is 1 foot (0.3 m), then it has a mechanical advantage of 3.0. We can lift (output) 3 times (300%) as much as the input weight.

That is just what a GHP does!

Chapter 20

Measurement of GHP Efficiency

This chapter goes beyond the basic definition of efficiencies to include how market-place competition raises COP and EER numbers, and how reducing electrical running costs increases efficiencies. The last section proposes a real-time continuous readout display of efficiency for the homeowner.

International, US and Canadian minimum efficiency standards have been established for heating in terms of the COP (coefficient of performance). In addition, standards are set for cooling, but defined slightly differently as an EER (energy efficiency ratio). The higher the COP or EER, the more efficient the system is. Both ratings measure efficiency using Output divided by the Input, but in slightly different ways.

Heating: COP (coefficient of performance) is the total Output heating capacity in Btu divided by the electrical power Input, also converted to Btu. Numbers range generally from 3 to 5, which translate to 300% and 500%.

Cooling: EER (energy efficiency ratio) is the total Output cooling capacity (in Btu/hour) divided by the power Input (in watts). Numbers can range from 9 to 31.

Homeowners must usually prove that they have met these efficiency standards in order to qualify for tax credits or rebates.

They can obtain certification from the manufacturer's website or from their installer. My contractor left a copy with the other documentation he provided.

GHPs are also rated by the ton. Most residential systems range from 2 to 5 tons. Mine is a 4-ton unit. A ton is a measure of the GHP's capacity to cool; it is the energy required to melt one ton of ice in 24 hours, which is 12,000 Btu per hour. (A Btu is defined as the amount of heat necessary to raise the temperature of one pound of water one degree Fahrenheit.) My system can move 48,000 Btu of heat energy per hour (at a specific earth-loop temperature) into my home. As the earth-loop temperature changes, so does the output. That is why it is so critical to have a proper earth-loop design to achieve maximum efficiencies. For example: the cooler the earth loop, the higher the output and efficiency when in the cooling mode.

Figure 20-1 shows the Energy Star current minimum heat pump efficiency standards compared to the current best available and to my open-loop system, installed in 2007. Government standards and tough competition are encouraging COP and EER numbers that exceed the minimums and so we are already seeing actual operating efficiency increase over time until a peak is reached. High efficiency can be a criterion for product selection in addition to reliability and warranties.

Comparison of Energy Star Standards vs. Real Systems						
System Type	Energy Star Minimum		Best Available		Author's GHP**	
	EER	COP	EER	COP	EER	COP
Closed Loop	≥ 17.1	≥ 3.6	34.3	5.2	--	--
Open Loop*	≥ 21.1	≥ 4.1	42.5	5.6	19.0	4.1
*Water-to-Air System ** Author's GHP installed in 2007 under lower minimums						

Figure 20-1

Eliminating COP/EER Confusion (Maybe)

Why are there two ways of measuring GHP efficiency in the US? In Europe, the term COP is used for both heating and cooling.

The COP (coefficient of performance) measures efficiency in the heating mode. The EER (energy efficiency ratio) measures efficiency when operating in a cooling mode. What is really confusing is that the COP and the EER use different energy units for computing the efficiency numbers. The COP uses Btu as the measurement unit of energy in both the numerator (output) and the denominator (input). That is really handy because a result of 3.3, for example, means that the efficiency is 330%.

The EER, however, uses Btu/hour in the numerator (output) and watts in the denominator (input). The result is a much larger efficiency number. To put it in COP terms, we must divide the EER number by 3.412 (converting watts to Btu/hour). Instead of an EER of 13, you get 3.8 or 380% efficiency.

Why the difference? Possibly because EER numbers were designed after the SEER (Seasonal Energy Efficiency Ratio) used for AC systems. Because GHPs *move* energy and because they are used year-round, they call it EER. In any case, when comparing companies, the higher the COP and EER, the better.

Increasing the COP in Installed Systems

There is a steady trend toward reducing the electrical costs of running GHPs. Scroll compressors replaced piston types, two-stage compressors replaced single-stage, infinitely variable stages are now shipping and possibly solid-state compressors are in the future. The impetus is two-fold: lower cost to run and higher COPs.

How does lowering the electrical input change the COP? If our starting equation is:

$$COP = \frac{\textbf{1.0}\ (electricity) + \textbf{3.0}\ (ground\ energy)}{\textbf{1.0}\ (electricity)} \approx \textbf{4.0}\ (or\ 400\%)$$

And we lower the Input electrical cost by only 10%, our COP goes from 400% to 430%; well worth the effort.

$$COP = \frac{\textbf{0.9} + \textbf{3.0}}{\textbf{0.9}} = \textbf{4.3}\ (or\ 430\%)$$

COP Measurement of Installed Systems

The COP number you would get in your home's installed system will be lower than the factory measurement because the factory cannot duplicate your specific installation. In addition, your actual COP will vary throughout the year based on several variables, such as colder weather pushing the system to the second stage, changes in the input ground-loop water temperature, or changes to the flow rate, to give just a few examples.

Actual measurement of COP in the home is fairly standard. It requires some equipment, expertise and knowledge of the system, but it is also rather cumbersome and only produces a single result at that point in time. It is not a continuous, real-time measurement. A suggested approach for real-time will follow for comparison.

We "just" fill in the formula previously shown:

$$COP = \frac{\textit{Input Electrical Energy} + \textit{Ground-Source Heat Energy}}{\textit{Input Electrical Energy}}$$

Equipment is needed to take five separate measurements: 2 temperature readings, flow rate, amperage and voltage:

Ground Source Heat Energy
1. Temperature of the ground loop Entering Water (EWT)
2. Temperature of the ground loop Leaving Water (LWT)
3. Flow rate of the ground loop water in gallons per minute (GPM); converted to gallons per hour (x 60).

(EWT – LWT) x GPM x 60 x 8.35 lbs/gal =   Ground-Source Heat Energy (Btu)

Input Electrical Energy
4. Amperage at that moment (A)
5. Voltage (V)

Note: 0.85 is a power factor, and 3.412 converts kWh to Btu.

A x V x 0.85 x 3.412 = Input Electrical Energy in Btu

An example:
1. EWT = 55°F
2. LWT = 48°F
3. GPM = 9

(55 – 48) x 9 x 60 x 8.35 = 31,563 Btu Ground Source Energy

4. A = 13.4
5. V = 230

13.4 x 230 x 0.85 x 3.412 = 8,938 Btu Electrical Energy

$$COP = \frac{8,938 + 31,563}{8,938} = 4.5 \ (or \ 450\%)$$

These rather cumbersome calculations give a COP of an installed system as a one-time measurement, not a continuous readout.

The Ultimate Goal: Real-time Continuous COP

My car's miles-per-gallon (MPG) display is useful because it encourages me to adjust my driving habits to get a higher number. There is currently no way to obtain a continual, real-time COP measurement of a heat pump in my home. Having such a meter is not essential, but would be very helpful in assessing the health of the system and in reducing the costs of running the system by changing our operating habits.

What the industry and homeowners need is a real-time, continuous display of COP that requires no clamp ammeter, no voltmeter or flow measurements. Ideally it would provide a continuous COP presentation on the heat pump console, or on your thermostat, your smart phone or on your computer. It would be similar to a MPG meter on a car.

Well, there is a way. This concept requires only two temperature measurements, both in the refrigerant loop: one at the entrance to the evaporator heat exchanger, T(cold), and the second at the entrance to the condenser heat exchanger T(hot).

Using the heating formula for a **water-to-air system**:

$$COP = \frac{T(hot)}{T(hot) - T(cold)} = 4.5 \ (or \ 450\%)$$

If we insert the temperature data from the schematic of the refrigerant loop (Figure 20-2), we must first convert those temperatures to Kelvin:

Evaporator   T(cold) = 28°F = 271°K
Condenser   T(hot) = 165°F = 347°K

$$COP = \frac{347}{347 - 271} = 4.5 \ (or \ 450\%)$$

EER monitoring in the cooling mode would be similar except that a factor of 3.412 must be added to compensate for the fact that the EER definition uses Btu/hour in the numerator (Input) and watts in the denominator (Output):

$$EER = \frac{T(cold)}{T(hot) - T(cold)} \quad \frac{271}{347 - 271} \quad x\ 3.412 = 12.1$$

For a **water-to-water system**, the COP works out to 5.3, showing that it is more efficient because it operates at 120°F. A lower temperature is accomplished by lowering the compressor pressure, which in turn reduces the electric power needed, and that increases the COP.

These calculations are based on an ideal process that is not quite possible in practice because it does not consider some of the internal losses. Actual numbers will be a bit lower.

The manufacturer and the wall meter can only measure the system's efficiency up to a certain point in the system: the final heat exchanger. A good portion of the system efficiency depends on the efficiency of the home's heat distribution system, the quality of GHP installation, losses in the ground loop, and even the temperature of the water entering the heat pump.

The factory numbers are still very useful. They show that they meet or exceed required standards and they permit comparisons between different units. The wall meter will help you set your system for optimal efficiency, much as an automobile MPG meter can help you achieve the most miles per gallon from your driving. ✪

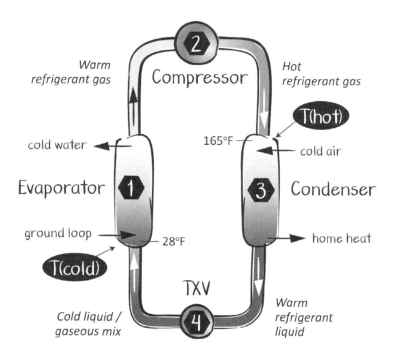

Figure 20-2

Real-Time, Continuous COP Measurement

For every 100,000 units of typically sized residential geothermal heat pumps installed, more than 37.5 trillion Btu's of energy used for space conditioning and water heating can be saved, corresponding to an emissions reduction of about 2.18 million metric tons of carbon equivalents, and cost savings to consumers of about $750 million over the 20-year-life of the equipment. *Source: EPA*

Chapter 21

The Importance of GHP Performance Standards

If you think that industry standards are only something for design engineers or manufacturers, but not really of interest to the homeowner, think again. Standards are a way of insuring that there is uniformity in how competing products are rated, measured, tested and (especially) compared. Standards are also in place for GHP design and installation; inadequate installation of heat pumps is one of the problems facing the industry.

Standards are for consumer protection. Without standards, companies would define product performance differently, test their products differently, and no one could compare one product to another. Mature industries want standards so they can design to meet or exceed them. Once a standard is set, testing methods can be devised for independent organizations to certify compliance of the product to those standards, insuring that ratings are uniform and that there is a fair comparison. Certification is voluntary, but competition is the driving force.

Of the many types of standards that apply to GHP systems—performance, quality, safety and electrical standards—this chapter will concentrate on performance standards.

Performance standards in the GHP field relate to efficiency standards. As discussed in earlier chapters, efficiency is the primary reason that GHPs are so much better than any other heating and cooling approach. Performance testing and ratings are done in a lab or factory where they measure the energy output at the heat exchanger. Therefore the results for homeowners will be lower because there is no way to test for onsite variables, such as ductwork losses, that are unique to every installation. These measurements are really conducted at the second stage, the refrigerant loop, and cannot predict a specific ground loop or heat loop.

Who Sets Standards?

ISO (International Standards Institute) is a non-governmental organization that forms a bridge between the public and private sectors. ISO has a network of national standards institutes in 159 countries, one member per country, plus members from national industry associations. Certification to ISO standards by third-party institutions allows the use of the ISO label. The ISO 13256 is the international standard for open-loop and closed-loop GHP systems. It basically lays out standard testing conditions and requirements.

Energy Star is a US government program, although it has also been adopted by Canada with a few changes to reflect their more severe weather conditions. The US Environmental Protection Agency along with the Department of Energy have set up an Energy Star program that includes GHP systems and their efficiency performance standards. It specifies minimum COP and EER ratings. Energy Star incorporates ISO 13256 and AHRI 870 *(see below)* into the requirements. The reward for agreeing to meet these standards is the right to display the Energy Star label. EPA's intention is to utilize the AHRI Directory of Certified Products to determine which equipment qualifies for Energy Star.

AHRI (Air Conditioning, Heating and Refrigeration Institute) is a US trade association representing over 300 manufacturers. It establishes industry standards and offers independent testing and certification to those standards by independent laboratories under AHRI contract. It is a voluntary, non-profit organization representing the major voice of the heating, ventilation, AC, refrigeration, and GHP industry. It established the original US GHP standards that were closely adopted by the ISO as an international standard. They have also established the AHRI 870 standard that is now the DX standard in the United States.

ANSI (American National Standards Institute) is the non-profit voice of US standards and conformity assessment. It adopts standards and provides accreditation programs to assess conformance with the standards. Headquartered in Washington, DC, ANSI is the official US representative to the International Organization for Standardization (ISO).

Energy Star Criteria		
Energy-Efficiency Criteria for Qualified Geothermal Heat Pumps		
Effective January 2013		
Product Type	**EER**	**COP**
Closed Loop Water-to-Air	17.1	3.6
Open Loop Water-to-Air	21.1	4.1
Closed Loop Water-to-Water	16.1	3.1
Open Loop Water-to-Water	20.1	3.5
DGX	16.0	3.6

Figure 21-1

IGSHPA (International Ground Source Heat Pump Association) is a non-profit organization established in 1987 to advance ground-source heat pump technology on local, state, national, and international levels. IGSHPA utilizes state-of-the-art facilities for conducting GSHP system installation training and geothermal research. Recently IGSHPA partnered with several industry associations to produce the National Certification Standard for Ground Source Heat Pump Personnel.

CSA (Canadian Standards Association) is a not-for-profit membership-based association serving business, industry, government and consumers in Canada. They adopted the ISO 13256 standard as C13256 for open- and closed-loop systems, while recognizing that their colder climate requires some tailoring to the standards. Those companies who meet the standards are permitted to display the CSA label. In addition, CSA standard C222 covers DX systems. Standard C448 covers the complete installation of open- and closed-loop systems: pipe quality, pipe fusion, foundation frost protection, ducting, drilling, grouting, etc.

Natural Resources Canada, Office of Energy Efficiency (OEE), is a center of excellence for energy conservation, efficiency and alternative fuels. It offers grants and incentives, workshops, statistics and analysis, and various publications. The Canadian Energy Star label is slightly different from the US label, displaying a bilingual "Energy Star High Efficiency—*Haute Efficacite*." ✪

The Geothermal Exchange Organization (GEO) is involved in the legislative and public promotion of industry standards. GEO is the voice of the GHP industry in the US. Existing tax and subsidy laws were the result of their efforts. *www.geoexchange.org*

Part III

The Broad View of Geo Power and Its Opportunities

Chapter 22

The Geothermal Marketplace

If I were an investor (I can only dream), I would look very closely at Canadian and US geothermal manufacturing companies. This is a growth market that will be pushed upward by the new tax incentives in both countries and by homeowners' desires to reduce living expenses.

Looking at the past not only provides some perspective about the future, but it also shows that the geothermal industry has over 60 years of operating experience. The first United States geothermal heat-pump system was installed in the Commonwealth Building in downtown Portland, Oregon, in 1946. This was a successful and highly publicized project that led to a number of installations throughout the country.

But widespread acceptance by architects, engineering firms and builders was very slow. The low cost of fuel and the high cost of the ground loops were barriers. There was a brief growth period during the oil crisis of the 1970s, but it was not until the last few years that commercial, institutional and residential applications have surged ahead as oil costs have increased. Ground-source geothermal heat pumps are now one of the fastest growing applications of renewable energy in the world. In

2008, Canadian heat pump sales increased by over 50%; even more in Ontario. US shipments increased by 40% in 2008 to 121,242 units.

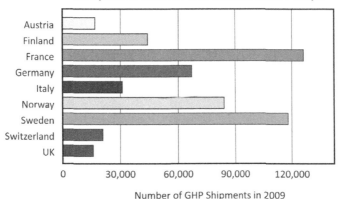

European Market of Geothermal Heat Pumps

Number of GHP Shipments in 2009

Figure 22-1

Source: *European Heat Pump Association; "Outlook 2010" Report*

However, this technology is currently only a small segment of the overall potential market. Total United States and Canadian new home starts in 2008 totaled over 1.1 million, and that does not include new building starts.

The Oak Ridge National Laboratory 2008 ground source geothermal report (see appendix, *Helpful Links and References*) concluded that although the United States has an installed base of about 600,000 units, the European markets absorb two to three times the number of GHP units per year than the United States. It is a worldwide growth industry, with rates in Europe and parts of Asia and Canada exceeding those in the United States. Sweden and France are the largest heat pump markets in Europe. In 2009, 524,000 units were sold in Europe; quite a jump from 92,000 in 2004.

The geothermal market is a part of the Heating, Ventilation and Air Conditioning industry, commonly referred to in the yellow pages as HVAC. The 2014 total HVAC market in the U.S. was around $70 billion.

Ground-source GHPs are a $2.5 billion dollar industry in the US that is growing by 30–40% per year as oil prices have surged. In the United States 36,439 GHP units were shipped in 2003, 83,396 in 2007, and 115,442 in 2009 for residential, commercial and institutional applications. So you can see that current GHP shipments are a small percentage of the total potential market. The industry's goal is to capture 30% of that heating and air conditioning market by 2030.

The current growth is producing growing pains, especially in terms of having enough qualified installers, drillers, and even sufficient materials, such as high-density polyethylene. But growth in the geothermal market involves a large number of small businesses, which translates to additional benefits for the local economies. It is a win-win situation for everyone. ✪

Growth of Geothermal Heat Pumps in the US Market

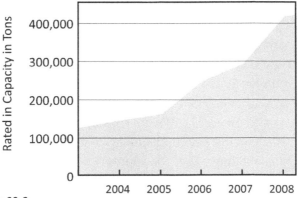

Figure 22-2

Source: *Energy Information Administration, "Annual GHP Manufacturers Survey" Report*

Commercial / Institutional Geothermal Systems

While this book is primarily directed toward residential applications, there is a growing and parallel effort underway for applying ground-based geothermal technology to a wide variety of government, institutional and commercial buildings, and even non-building applications. What made this so interesting to me was the amazing diversity of applications; applications that I had never considered.

Since projects of this type are usually larger in terms of energy, they have a larger impact on energy use. GHP systems provide greater savings for commercial buildings with their large heating and cooling load. The payback period is usually four to eight years, but it could be as short as two to three years. The first United States geothermal heat pump system installed in 1946 was a commercial installation.

Today's commercial GHPs are considered a cost effective, energy efficient and environmentally friendly way of heating and cooling buildings. They are appropriate for new construction as well as retrofits of older buildings. Their flexible design requirements make them a good choice for schools, churches, high-rises, government buildings, apartments and restaurants.

Commercial installations also have significant advantages with geothermal:

▸ Different parts of the building can be simultaneously heated and cooled. For example, the sunny side may need cooling while those on the shady side need a bit of heat.
▸ The quietness of operation is a benefit to the building's occupants; they won't even know when the system is running.
▸ Multiple zones allow individual room control.
▸ Less mechanical space is required (50–80% less!).
▸ No outside equipment to hide (and no rooftop units), which also eliminates vandalism.
▸ All-electric system eliminates multiple utility services.
▸ No boiler and chiller maintenance.
▸ Lower life-cycle costs.
▸ Lower peak demand and lower operating costs.
▸ Big savings on energy consumption, from 25–50%.
▸ Water for consumption can be heated with waste heat in the summer at no cost, plus reducing costs in winter.
▸ Rebates and tax incentives available.

Canada, with cooler weather and lower average temperatures which results in higher heating costs, has been adept at finding innovative applications for ground-source geothermal technology, including fish hatcheries, tree nurseries and curling rinks.

Agricultural and horticultural businesses also use geothermal energy in a variety of applications. For example, dairy farms use GHPs for milk pasteurization and chilling as well as hot washing of equipment. Greenhouses in several western states use direct thermal wells for heating, while New England and Canadian greenhouses are successfully applying heat pump technology for this application. Large-acreage tree nurseries are typical examples where fuel savings of over $10,000 per year are being obtained. Here are just a few other examples that show the wide range of geothermal applications:

Smart Bridges—Research is underway in Oregon and Oklahoma to use GHPs to circulate heated fluid through tubes embedded in bridge decks to reduce icing, which in turn, will eliminate salt usage and the resulting corrosion. These 'smart' bridges are tied into a network of weather stations to predict icing conditions.

Gas Stations—One major gas station/convenience store chain is retrofitting one store per week with geothermal systems. These systems absorb waste heat from freezers and icemakers and even provide hot water for a car wash area.

Government Buildings—At the west entrance of the Colorado State Capitol in Denver, the fifteenth step has a plaque engraved "One Mile Above Sea Level." In 2010, deep beneath those steps, two 900-foot (274-m) holes were drilled for an open-loop, $6 million geothermal heating and cooling installation that will provide an estimated savings in utility bills of $95,000 per year for the state. The supply well taps into the 55°F (13°C) ground water heat of the Arapahoe aquifer; the injection well receives the water after the heat has been extracted.

Hotels—One of the largest GHP installations in the United States is the Galt House East Hotel in Louisville, Kentucky. Heating and cooling are provided for 600 hotel rooms, 100 apartments and almost a million square feet of office space. The hotel saves $25,000 per month compared to a similar non-GHP building adjacent to it.

Commercial Profile: Baxter Building

The Baxter Building is a 14,000 square-foot (1,300 sq m), mixed-use office building in Poughkeepsie, New York, that proves DX GHP systems are equally applicable to commercial buildings. It uses a DX geothermal heating and cooling system because the cost of energy for a traditional HVAC—which would've required over 4,000 gallons (15,141 liters) of oil each year for heating—made the property owners and developers look for alternatives.

The builder, R.L. Baxter, received geothermal bids from two companies. One bid specified a closed-loop system using deep wells with over 8,000 feet (2,438 m) of cased well drilling. The other bid from Total Green Geothermal of Monroe, New York, was accepted. They offered a DX approach with higher conductivity, more efficient copper piping and refrigerant in the ground loop. Six DX fields were placed in the parking lot and covered with environmentally friendly asphalt that allows drainage through the surface (this moisture actually allows the fields to perform better by improving heat conductivity). Fifty-six holes, each 70 feet (21 m) deep, were drilled at 30-degree angles. Advanced Geothermal Technology from Reading, Pennsylvania supplied four 5-ton and two 4-ton DX units.

Using green technology not only saves money, but assists in the marketing of such buildings.

Schools & Colleges—Recently the US Department of Energy reported that over 500 schools in the United States have installed geothermal heating and cooling systems. The ability to cool school buildings permits year-round occupancy, if needed. The Comanche Elementary School in Comanche, Oklahoma has an interesting GHP installation because it uses the city water supply as its heat source to get very energy-efficient heating and cooling. This idea appears to open new city applications where trenching is not an option.

The Australian Outreach College in Brisbane, Australia has installed a direct exchange (DX) geothermal system to cool classrooms and an integrated- technology building that has high heat loads from people and computer equipment.

Bard College in New York State has built a set of dormitories heated and cooled with a GHP using vertical boreholes.

Hospitals—The new Sherman Hospital Medical Center near Chicago uses a 15-acre (6-hectare) lake as a heat source for heating and cooling, claiming a savings of over $1 million a year in energy costs. The hospital received a $400,000 grant from the State of Illinois and $956,000 from DOE.

Airports—In 2009, as part of a major upgrade, the Juneau, Alaska, airport installed a ground-source geothermal heat pump system that is expected to reduce energy costs by $85,000 per year. That same year, the Nantucket Memorial Airport completed a similar system.

Churches—In the US, Canada and throughout Europe you can find stories of both large and small, very old and very new churches being heated and cooled using heat pump technology. An interesting twist occasionally resulted. In retrofit examples where insulation was inadequate, the general results show that the annual heating costs were not lowered after a GHP installation. What was gained, however, was cooling at no extra cost, providing a higher level of comfort and increased attendance and contributions in the summer. As temperatures generally increase around the world, this will become even more important.

Trinity Church, Boston, MA

In the late 1980s, my wife and I decided to experience city life and what better city to choose than Boston? We bought a newly renovated 1850-era South End condominium where we had the first floor, lower level and a rear garden. It was within walking distance of Martha's art studio in an ancient former distillery building. It was also close to the Charles River, the Esplanade, the Public Gardens and the golden-domed State House for walking and bike riding.

And then there was Copley Square where, when standing at the Boston Public Library, you could see across the plaza to one of the most beautiful churches in the United States, Trinity Church—a Boston historic landmark built in 1877. Although not a member, I often stopped in to see the sunlight through the stained glass windows, the sculpture by Saint-Gaudens, the murals, or just sit and enjoy the quiet atmosphere.

That Back Bay area consisted of filled, marshy wetlands in the 19th century so that the architect, H.H. Richardson, had to build the massive 9,500-ton church on four large granite pyramid-shaped piers that sat atop thousands of wood pilings. After 125 years, a great number of repairs were needed. And, because of a changing water table, many of the exposed pilings were deteriorating, threatening the integrity of the building.

After years of planning, a complicated and massive restoration/expansion project was initiated in 2001. The design and engineering teams struggled with where to put the new mechanical systems for the expansion. The church's steep roofs and spires prohibited a typical roof-mounted cooling system. Instead, they chose a geothermal energy system. Six wells, each 1,500 feet (457 m) deep, were drilled within feet of the structure. Thirteen heat pumps totalling 130 tons of capacity were installed. Estimated savings are an impressive $67,000 each year.

Australian Government Science Building—The Geoscience Australia building in Symonston, Australia, is the largest ground-source geothermal installation in the southern hemisphere (and maybe the world). It uses 210 heat pumps and 352 boreholes, each 320 feet (98 m) deep. It has been operating for 10 years.

Canadian Steel Plant Sludge Drying—To dry sludge for more economic transfer to a landfill, a closed-loop geothermal heat pump is used to heat and dry air that is passed through the sludge to pick up moisture which is then extracted by a condenser.

Seawater-Source Heat Pump—A 28,000 square-foot (2,600 sq m) research center in Norway has installed a geothermal heat pump that uses seawater as the heat source and ammonia as the refrigerant.

Tree Nursery—A New Brunswick, Canada nursery uses four 35-ton GHP units for greenhouse heating.

Fish Farm—A fish hatchery in Canada employs a GHP to raise the incoming fresh water temperature for optimal hatching and growth.

Civic Center—The Port Hawkesbury Civic Center in Nova Scotia, Canada, is a multi-purpose building of 22,660 square meters (244,000 sq ft) that uses a geothermal heat pump. An Ice Kube geothermal chiller cools the ice rink and uses the heat for radiant floor heat, melting of sidewalk snow, and melting of ice shavings from the rink. Forced air heat pumps also provide heating and cooling. ✪

Chapter 24

Future Trends

As with most technologies, competition forces companies to invent improvements to gain a market edge. Many interesting new trends are in the works.

Increasing GHP Efficiency

The unusually large efficiency of a GHP system is what makes it such a powerful candidate for home or building heating/cooling. But the GHP level of efficiency has not yet peaked. In the coming years, I expect it will increase past 600%, perhaps even 700%, as engineers find ways to extract more heat energy out of the ground and components are improved to do the same amount of work using less electric power.

Direct Financing

Contractors and manufacturers are starting to offer 100% financing so that the system owner is not tasked to find bank loans.

Geothermal Leasing

Electric utilities in Minnesota and Colorado are now installing, paying for and leasing back the ground loops for customers'

GHP systems for a fixed monthly tariff. This is smart, long-term thinking and a huge step in making GHPs more affordable. The business logic is that the ground loops can be considered a utility investment much like power poles and wiring. Indeed, with many people moving from existing homes in five years or less, the concept of leasing the entire GHP package is actually under consideration (or at least up to the ductwork, radiant floor or hot water heat delivery subsystem). This would eliminate the burden of financing and would probably be a package that includes all service, repairs and maintenance.

On-Board Telemetry

On-board telemetry will allow the owner to monitor the actual energy consumption on a daily or monthly basis to verify system benefits and control costs. It will be like the miles-per-gallon meter in your car. This concept can be expanded to calculate annual greenhouse emissions to aggregate credits for the cap-and-trade market where and if it goes into effect.

Variable-Speed Compressors

The first compressors had one speed no matter what size of load. Then two-stage units permitted the compressor to coast at a lower level to meet lower needs, saving electricity, while the second stage was available for higher loads. Expect to see infinitely variable stages that will automatically adjust to the exact level needed.

Solid-State Compressors

It all started with a piston-type mechanical compressor with over 20 moving parts requiring oil for lubrication, then converted to a higher-efficiency scroll compressor with only six moving parts and then to a digitally controlled scroll compressor. What is inevitable in the long term is a solid state device with no moving parts and no oil required. Research is also underway in magnetic pulse compressors, electrochemical approaches, superconducting cryogenic devices and several other concepts. It will first be applied to refrigerators and then to GHPs, but it will not be here next week.

Better Building Efficiency

The growth of geothermal and solar will increase the home-owner's interest in building efficiency and load reduction. There is an increasing trend for contractors to look at the total picture: house design, solar, geothermal and wind installations.

A zero-energy home (ZEH) is a general term applied to a building that has net zero energy consumption and zero carbon emissions annually. A ZEH combines high levels of energy efficiency with state-of-the-art renewable energy systems to annually return as much energy to the utility as it takes from the utility, resulting in a net-zero energy consumption for the home. A geothermal heat pump fits perfectly into net-zero and green building goals.

A "green building" is commonly defined as using architectural resources more efficiently and to reduce a building's impact on the environment. The LEED (Leadership in Energy and Environmental Design) system is an internationally recognized building certification program developed by the US Green Building Council (*www.usgbc.org*). It provides check lists as measurement tools to achieve green building goals that can be used for third party verification. It applies to both commercial and residential buildings.

Net-zero homes and green buildings can deliver one important goal: to completely or very significantly reduce energy consumption and greenhouse emissions for the life of the building. This involves the following disciplines:

▸ Don't build more space than you need.
▸ Build the most efficient building you can afford (the best windows and insulation, optimum orientation for passive solar heating, etc.).
▸ Add solar electricity and solar water heating to the mix.
▸ Reduce energy demand.

To learn more, visit the National Home Builders Association's Green Building Program: *www.nahbgreen.org*

Official Installation Standards

Standards for GHP hardware compliance are in use everywhere, including efficiency standards, but there are no officially approved national AHRI *installation* standards as of 2014. Until this is addressed, the International Ground Source Heat Pump Association (IGSHPA) provides an installation standard that is used for accreditation training of installers in the United States. Canada has adopted standard C448 as their installation standard.

Shortage of Qualified Installers

As incentives kick in and public awareness increases, the number of GHP requests will accelerate. A lack of qualified contractor/installers may become a problem. Here is where jobs will be looking for people. Accreditation courses are necessary but cannot make up for the lack of hands-on field experience and apprenticeship. ✪

Chapter 25

Geothermal Careers: Joining the Industry

If you are considering a career change, you should look carefully at the geothermal field, especially if you have some HVAC experience. It is growing field with a shortage of qualified personnel. High demand often means higher pay.

Most of these jobs are considered to be mid-level or middle-skill careers that demand more than a high-school diploma but less than a four-year degree. These well-paying jobs require training and usually accreditation.

A typical drilling rig. Photo courtesy of Lawrence A. Muhammad, Geo NetZero

Middle-skill jobs, which are critically important to our economies, have been disappearing for the last 30 years because of
technology changes and off-shoring. Compounding the loss of
jobs in this sector was the recent recession.

One bright spot, however, has been the growth of green jobs
and these jobs have grown 2.5 times faster than all other jobs
(Pew Charitable Trusts report, 2009). As a subset within that
green zone, GHP manufacturing shipments are still increasing
every year, aided in great part by large tax incentives. Every
single GHP shipment from the manufacturer requires a number
of local, trained experts in loop construction or borehole drilling,
someone to apply software programs to determine system size
and ground loop design, someone who understands the GHP
circuitry and controls and can hook it all together to make it
work. That could be you.

Geothermal Job Opportunities and Standards

The National Certification Standards for Ground Source Heat
Pump Personnel brings together for the first time an analysis of
the roles and responsibilities of each of the individual job tasks
involved in the design and installation of GHP systems. This
project, managed by the Geothermal Heat Pump Consortium
and in partnership with IGSHPA and many others, addresses
the following job opportunities in the geothermal industry:

▸ Ground Source Heat Pump System Project Manager
▸ Ground Source Heat Pump System Engineer/Designer
▸ Geological Formation Thermal Properties Tester
▸ Ground Source Heat Pump System Commissioning Agent
▸ Vertical Loop Driller
▸ Horizontal Directional Driller

‣ Ground Heat Exchanger Grouter
‣ Ground Heat Exchanger Looper
‣ Ground Source Heat Pump System Water Well Driller
‣ Ground Source Heat Pump System Water Well Pump Installer
‣ Ground Source Heat Pump Mechanical System Installer
‣ Ground Source Heat Pump System Operations/Maintenance Technician
‣ Ground Source Heat Pump System Inspector/Regulator
‣ Ground Source Heat Pump System Trainer

For more detailed information, visit *www.geoexchange.org* and search under Regulatory, or *www.osti.gov/scitech/biblio/1116539*

How to Join the Industry

Throughout this book, I have pointed out that no GHP system can be entirely successful without a high quality installation by trained professionals. The two career jobs needed most to do this are GHP System Installers and GHP Vertical Loop Installers. For a taste of the type of material covered by the training, note the simplified course listings on the next page.

Both of these jobs require US or Canadian certified training plus field experience. In most cases, the training and certification must come first. Where can you find a friendly, certified trainer in your area? There is no single database that lists all US and Canadian trainers or training organizations. But one comes close. The International Ground Source Heat Pump Association (IGSHPA) lists all IGSHPA accredited trainers in the US and Canada, their organizations, locations and contact information (*www.igsha.okstate.edu/directory*). Or search the internet for "training geothermal heat pumps North Dakota" or whatever state you are in.

You will find many organizations in the US and Canada provide GHP training and certification. It is impractical to list them all here, but I have provided a few samples to show the wide types of organizations that provide GHP training. These include dedicated training companies, GHP contractor/installer companies, GHP manufacturers, and very importantly, non-profit, industry-sponsored geothermal organizations. I have put these at the top of the list because they provide education and training as well as promote industry standards and set the level of professionalism for the industry.

I am particularly impressed with Canada's approach to professionalism in the GHP field. In Canada, training alone will not provide a certificate. Accreditation is not given until documented experience has also been demonstrated. That is a big difference that will benefit the homeowner, the industry and the installers (although not all of them may agree with me). In the US, a 3-day workshop provides an installer's card and certificate; job experience is up to the individual or company. ❂

A header system being installed.
Photo courtesy of Lawrence A. Muhammad, Geo NetZero

- IGSHPA Accredited -

GHP System Installer Training Subjects

DAY ONE
1   GeoExchange Introduction & Overview
2   Economics, Marketing & Demand Reduction
3   Soil and Rock Identification; Thermal Conductivity
4   Selecting, Sizing and Designing the Heat Pump System

DAY TWO
5   Designing the Ground Heat Exchanger
6   Grouting Procedures for Vertical Ground Heat Exchangers
7   Installing the Ground Heat Exchanger
8   Pipe Joining; Fusion

DAY THREE
9   Flushing, Purging and Flow Testing the Heat Pump System
10   Heat Pump System Start-up and Checkout
11   Project Costing and Bidding for GeoExchange Drilling
12   Review, Conclusions

Vertical Loop Installer Subjects

▸ GHP system design and layout basics
▸ System materials
▸ Pressure drop calculations
▸ Thermal conductivity
▸ Drilling processes
▸ Containment procedures
▸ Grouting concepts
▸ Air and debris purging
▸ Pipe joining techniques
▸ Project bidding
▸ Partnerships

Where to Find Training

In-Depth Training Manuals

IGSHPA
www.igshpa.okstate.edu
The International Ground Source Heat Pump Association provides the following training manuals. Check their website for more details.

Residential and Light Commercial Design and Installation Manual

Slinky™ Installation Guide

Design and Installation Standards

Soil and Rock Classification Field Manual

Grouting for Vertical GHP Systems

CLGS Installation Guide

Industry Organizations that Provide Training

IGSHPA
www.igshpa.okstate.edu
The International Ground Source Heat Pump Association, located in Stillwater, Oklahoma, is a non-profit, member-driven organization established in 1987 to advance geothermal heat pump technology at local, state, national and international levels. They conduct training on the campus of Oklahoma State University. Traveling workshops are also available. Their annual 3-day conference brings together experts from the US and Canada to share the latest information on ground-source systems. They also produce publications such as *Ground Source Installation Standards*. Successful completion of the training and the exam for the Accredited Installer Workshop or the Accredited Vertical Loop Installer course provides an IGSHPA installer's card, a certificate, a complete set of manuals and membership in the IGSHPA.

Canadian GeoExchange Coalition
www.geoexchange.ca
CGC is an industry-based, non-profit organization located in Montreal, Quebec, Canada. It provides qualification of firms, certification of products, plus training and accreditation of individuals. It offers 3-day workshops for installers and drillers which provides a CGC Training Certificate. To actually receive CGC installer accreditation, the applicant must also demonstrate field experience.

GeoExchange BC
www.geoexchangebc.ca
A non-profit industry association located in Surrey, British Columbia, Canada, that provides information, education and training for the heat pump industry. Their 2-year comprehensive Certified Geothermal Technician apprenticeship training program provides a Certificate of Qualification. For those with 10 or more years of experience in the industry, a transition period is available to earn a Certificate of Qualification without additional training. They also offer CGC-approved training workshops and certification for drillers and installers.

Iowa Geothermal Association
www.iowageothermal.com
A non-profit, industry-sponsored trade organization in Johnston, Iowa, with a goal of promoting and ensuring quality heat pump installations. They work with the Iowa Energy Center, a research and education organization, to provide IGSHPA GHP installer training and accreditation courses.

Specialized Training Companies

HeatSpring Learning Institute
www.heatspring.com
An education company located in Cambridge, Massachusetts, that focuses on clean energy training. They offer IGSHPA geothermal installer training as well as a 6-week online entry-level, non-IGSHPA training course.

Geo NetZero
www.GeoNetZero.com
A Detroit, Michigan, company that provides geothermal renewable and alternative energy technology accreditation workshops for the unemployed, underemployed, and workers returning to the workforce, as well as current workers needing workshops related to geothermal technology installations, commercial or residential.

GEOptimize
www.geoptimize.ca
A GHP design and consulting firm in Winnipeg, Manitoba, Canada, that provides Certified GeoExchange Designer training and tools for both residential and commercial applications.

Examples of Contractors & Manufacturers that Provide Training

Eagle Mountain
www.eagle-mt.com
Canandaigua, NY

Earth Energy Services
www.earthenergyservices.com
Milwaukee, Wisconsin

GeoSmart Energy
www.geosmartenergy.com
Cambridge, Ontario, Canada

Geothermal Training Institute
www.geotrainers.com
Maple Plain, Minnesota

Greenville Technical College
www.gvltec.edu/geothermal
Greenville, South Carolina

Major Geothermal
www.majorgeothermal.com
Wheat Ridge, Colorado

Nexus Energy Products
www.nexusenergyproducts.com
Morden, Manitoba, Canada

Frequently Asked Questions

How It Works

Can one geothermal system provide both heating and cooling for my home?
Yes. One of the advantages of a GHP is that instead of two separate systems, one push of a button on the thermostat switches from heating to cooling.

Can a GHP be installed in new homes as well as older homes?
Yes. GHP systems are being installed every day in new construction as well as in older homes.

What are the alternatives for the basic configurations of earth loops?
They fall into two categories: open loop and closed loop. **Open loops** take water from a pond or well, extract the heat and then return it to the same or another location. New water is constantly being used. **Horizontal closed loops** circulate water in plastic pipes placed in trenches 5–6 feet (1.5–1.8 m) underground and then return it to the GHP, extract the heat and then recycle the water back into the ground in a continuous operation using the same water. **Vertical closed loops** send the water into vertical boreholes that may be as deep as 70 feet (21 m) to perhaps 600 feet (182 m)

and then return it to the GHP. **DX or Direct Exchange** systems use copper piping instead of plastic and send the refrigerant through the ground and back. Instead of having a separate ground loop and refrigerant loop, they are combined into a single loop. See chapters 4 and 5 for details.

If I use my well in an open loop system, what capacity or flow rate must I have?
Your manufacturer will specify the flow rate needed for your system, but the rule of thumb is 2–3 gpm (7.6–11.3 lpm) per ton of GHP capacity. So, a 4-ton unit needs at least 8 gpm (30 lpm). Now, your domestic needs must be added in. Many people find that 2 gpm (7.5 lpm) is adequate.

What are the advantages and disadvantages of these various configurations?
Open loops are generally less expensive because there is less trenching. Horizontal closed loops require more available land space. Vertical boreholes require less space, but are more expensive to drill. DX systems are usually more efficient and require less ground space since the holes are drilled at angles from a central point. A sacrificial anode is used in acidic soils to prevent copper

deterioration. Chapter 7 has additional information.

How far apart are the trenches or vertical boreholes placed?

Ideally, vertical boreholes should be at least 15 feet (4.5 m) apart, preferably 20 feet (6 m) apart. The centers of each horizontal trench should be about 10 feet (3 m) apart.

Can I install my own geothermal system?

Yes, if you are a certified, trained and experienced installer. Otherwise, forget it. Read chapter 15 for why this is important.

How much space should I allow for the interior equipment?

The floor footprint of the heat pump is about the same as a refrigerator, but ductwork and piping can double that.

How can a GHP heat my domestic water?

A desuperheater is an optional add-on with a separate heat exchanger that can provide excess heat to your hot water tank. Chapter 10 describes this plus other methods of obtaining hot water.

Is a geothermal heat pump noisy?

No, a GHP is very quiet, especially compared to fossil fuel burners with their abrupt, noisy start up while sucking in huge amounts of air.

Are GHPs safe?

Well, they are a little hard to steal. Sorry, could not resist that. The lack of any combustion means no flames, no noxious gases. They are as safe as your refrigerator.

Can a GHP be connected to an existing radiant system?

Absolutely, and also to an existing hot water system, or a hot air system. This can save considerable expense. Radiant or hot water will not permit heat pump air conditioning, of course. For hot air, you should check with your contractor. Many fossil fuel systems have a high velocity air flow that use smaller-sized ductwork. A GHP saves money by using a slower, more comfortable rate of flow, but requires a larger sized duct.

Will an underground horizontal loop affect my landscaping and lawn?

With horizontal trenches, yes, but only for a week or so. The grounds will soon be back to normal. It is also possible to bore underground horizontally if your contractor has the proper equipment. This reduces the impacted area considerably. See chapter 4.

What level of maintenance is required?

Very little maintenance is needed. There is no outside equipment, as with an air conditioner; no annual tune-ups or nozzle replacement, as with an oil burner; and no chimney to clean. Air filters need replacement as with any air delivery system. Water filters need cleaning if you have an open-loop source.

Do I need a programmable thermostat with my GHP?

Not really, but it will not come without one. You gain a lot of flexibility, plus the installer needs it to set a number of important parameters. A GHP thermostat does a lot more than just set temperatures. For example, with the push of a button (or automatic programming) it switches from heating to cooling mode. For maximum efficiency, experts recommend that you set the GHP thermostat and never change it. Each zone will come with its own thermostat. You'll find the programming capability really quite handy.

Do systems in cold climates need additional heat sources?

While properly installed GHP systems will provide all the heat or cooling needed, all systems require an emergency backup system. For a possible compressor failure, there is a backup electric subsystem built-in. For a power failure, the GHP is in the same unhappy position as any standard system. A backup emergency device is very comforting.

The Nuts & Bolts of the Technology

How do you transfer heat energy from the earth?

The earth has the ability to absorb and store heat energy from the sun. If we want to use that stored earth energy, we can extract heat through a liquid medium that is colder than the earth temperature and then transfer that heat energy to the heat pump. Heat will always flow spontaneously from a warmer medium to a cooler medium.

Will a GHP lower the earth temperature if we keep pulling out heat energy?

No, the earth absorbs 47% of the sun's energy that reaches the earth. This is 500 times the total energy needs of the planet. Locally, the ground temperature can be modified temporarily, merely reducing the efficiency a bit. But during summer cooling, the heat energy is returned to the earth.

I put my hand in 50°F water and it was very cold! How can that possibly heat my home?

Any material at 50°F contains a large amount of heat energy. It feels cold because your hand temperature is near 98.6°F. Heat energy is being transferred from your hand to the water, warming the water. And that is because Mother Nature insures that heat energy spontaneously flows from the warmer to the cooler object. But assume for a minute that you remove your hand and insert a pipe containing a liquid at 40°F. The heat in that 50-degree water would be transferred to the cooler refrigerant.

But I still do not see how that 50°F water ends up providing 72°F room temperature if I'm not burning anything.
Actually, that 50-degree water ends up providing much higher temperatures than 72 degrees. To make a room feel comfortable, we must provide hot air that is higher than skin temperature, or about 100–105°F. To reach that temperature, we must pass the air over a heat exchanger that is about 165°F to compensate for the heat it loses as it passes through many feet of cooler metal ductwork. By using a compressor and a series of valves and controls, your geothermal energy system works its magic to give us very high temperatures for home heating. See chapter 16 for a detailed explanation of what goes on inside the GHP.

How does a geothermal heat pump cool the home?
By pushing a button on the thermostat, you can immediately switch from Heat to Cool. This activates a reversing valve that changes the flow of heat energy from your home into the ground. **Why does the geothermal**

ground loop need antifreeze when it is below the frost line?
The water entering the ground loop often drops below 30°F (1°C) after transferring its heat. The lower the temperature, the better the heat transfer from the ground. Heat transfer requires a temperature differential; the greater the difference, the more efficient the transfer.

Are we wasting water with an open-loop system?
No, we are just returning it back to the aquifer where it came from.

If I use a water softener, should the open-loop geothermal water also be softened?
No, do not condition water for the GHP or for water used for outdoor purposes.

How do I determine how efficient my GHP unit is?
Look for the Energy Star label or go to your manufacturer's web site. The coefficient of performance (COP) indicates heating efficiency. A closed-loop system should have a COP rating of 3.6 or higher; an open-loop system should be 4.1 or higher. Chapter 19 provides more information, including definitions about GHP efficiencies.

Savings / Costs / Warranties

How much does a GHP cost?

Buying a heating/cooling GHP system is not like buying a refrigerator off the shelf. Each installation is unique and costing requires a qualified contractor. Chapter 12 provides more details. Keep in mind that extra costs include piping, ductwork, and ground trenching or drilling. Visit manufacturers' websites for their online savings/cost calculators.

Will a GHP reduce my living expenses?

A GHP will significantly cut your annual heating/cooling costs every year. In addition, the costs to maintain the system will be lower.

What financial incentives are provided?

Substantial US federal incentives provide a tax credit of 30% of the cost for GHP systems, but these tax credits are set to expire at the end of 2016, unless the government votes for an extension. Your state or power company incentives can to found online at the Database of State Incentives for Renewable Energy (*www.dsire-usa.org*). Canadian incentives are changing in 2010 (*www.oee.nra.gc.ca*).

What is a typical payback period for a GHP system?

Payback is defined as the time in years that it takes to recover the added costs of a GHP compared to a standard heating plus cooling system. It can range from zero to 5 years or more. Chapter 13 explains more.

Does a GHP increase the value of my home?

Yes, it does; the EPA has stated that it increases $20 for every dollar of energy saved. If you save $2,000 per year by not buying fuel oil or natural gas, the increase in your home's value will be $40,000. Whether that is real or not depends on the housing market and the buyer, of course. Please refer to chapter 8 for a more rigorous discussion of this.

How long will a geothermal system last?

While a standard oil burner has a 12–15 year life, a geothermal heat pump can last 20–25 years. An oil burner has large bursts of energy followed by sudden, noisy air intakes and chimney exhausts that stress the system. GHPs have smaller changes, fewer working parts, and do not combust anything. Refrigerators using the same heat pump technology can last 30 years or more.

Do geothermal systems have warranties?

GHP manufacturers offer equipment warranties, varying from 1 to 5 years, or more. This is one criteria for judging/selecting a manufacturer. The appendix lists manufacturers in the US and Canada along with their warranties. You should also get an installation warranty from your local installer.

Environmental Benefits

How does geothermal energy benefit our environment?

All other HVAC methods use up energy sources that are not renewable or produce pollutants into the atmosphere. A GHP emits no pollutants on-site and uses energy in the earth that is continually renewed by nature. GHP systems are the lowest emitter of CO_2 and the best decarbonization system for heating and cooling.

Helpful Links & Resources

US & Canadian Government Agencies

www.dsireusa.org
Database of US State Incentives for Renewables & Efficiency is a comprehensive source of information on state, local, utility and federal incentives and policies that promote renewable energy and energy efficiency.

www.eere.energy.gov/ geothermal
US Department of Energy, Energy Efficiency and Renewable Energy website includes geothermal information and education for consumers, DOE blog to ask questions, database on funded programs, and incentives.

www.energy.gov
US Department of Energy

www.energystar.gov
US EPA listing of all companies (including heat pump manufacturers) who are Energy Star Partners, plus much more.

http://oee.nrcan.gc.ca
Office of Energy Efficiency (OEE) of Canada manages the ecoEnergy Efficiency Initiative, Center of Excellence for energy conservation and efficiency. Lists Canadian federal, provincial, territorial and municipal incentive programs.

Geothermal Heat Pump Organizations and Websites

www.ahri.org
American Air Conditioning, Heating and Refrigeration Institute(AHRI) certifies heating, AC and geothermal efficiencies; has a database of certified companies and products.

http://cgec.ucdavis.edu
California Geothermal Energy Collaborative addresses large-scale geothermal power as well as GHPs.

www.csa.ca
The Canadian Standards Association (CSA) is a non-profit membership based association that develops standards, tests and certifies products for Canada and US markets and trains and accredits personnel.

www.earthcomfort.com
Michigan Geothermal Energy Association

www.geoexchange.ca
Canadian GeoExchange Coalition (CGC) promotes public awareness, training, accreditation and certification to CSA standards.

www.geoexchange.org
Geothermal Exchange Organization is a non-profit, trade association sponsored by the Geothermal Heat Pump Consortium for the geothermal heat pump industry. They educate consumers and professionals with online certification programs/workshops, provide a comprehensive GeoExchange Directory of contractors in the US and Canada, and conduct GeoExchange Forum online to discuss geothermal issues.

www.geoexchangebc.ca
GeoExchange BC, a non-profit industry driven organization located in British Columbia, provides education, training and certification in association with the CGC.

http://geoheat.oit.edu
The Geo-Heat Center at the Oregon Institute of Technology Center in Klamath Falls conducts certification and training courses in geothermal and solar PV technology and provides information, case studies, workshops, technical papers, etc.

http://geothermal.marin.org
Geothermal Education Office

www.greenbuildingtalk.com
Find homeowner questions, reports on experiences, info on heat pumps, comments on companies, etc.

www.gogeonow.org
Colorado Geo Energy Heat Pump Association

www.heatpumpcentre.org
International Energy Agency's information service to accelerate use of heat pumps, conferences, newsletters, workshops, etc.; based in Sweden.

www.heatspring.com
The HeatSpring Learning Institute in Cambridge, Massachusetts, is an education company providing clean energy training in geothermal heat pump and solar systems.

www.hvac-for-beginners. com/geothermal-ratings
A site that ranks GHP companies, mostly by warranty.

www.igshpa.okstate.edu
The International Ground Source Heat Pump Association is a non-profit organization established to advance GHP technology through technical conferences, a database of state incentives, geothermal research, and installation training.

www.iowageothermal.org
Iowa Geothermal Association

www.minnesotageothermal heatpumpassociation.com
Minnesota Geothermal Heat Pump Association

www.nahbgreen.org
National Green Building Program by the National Home Builders Association

www.renewables.ca
Online newsletter on renewable energy activity in Canada.

www.wisgeo.org
Wisconsin Geothermal Association

Other Reference Material

Geothermal Heat Pumps: A Guide for Planning & Installing by Karl Oschner, *www.earthscan.co.uk*

Got Sun? Go Solar 2nd Edition by Rex Ewing and Doug Pratt, *www.PixyJackPress.com*

Ground Source Heat Pump Analysis published by RETScreen International Clean Energy Support Center of Canada (*www.retscreen.net);* an electronic textbook for professionals and university students.

An Information Survival Kit For the Prospective Geothermal Heat Pump Owner by Kevin Rafferty, PE, distributed by HeatSpring Learning Institute; *www.heatspring.com*

National Certification Standards for Ground Source Heat Pump Personnel
Identifies job tasks, education, qualifications, experience, and skills required. Look under Regulatory at *www. geoexchange.org*

Residential GHP Manufacturers

as of October 2014

The following is a list of US and Canadian manufacturers of residential ground-source geothermal heat pumps. I've made no attempt to judge their quality or performance. For up-to-date information and to find new companies, please check the internet.

All of these companies offer scroll compressors and most have dealer locators on their websites. Whether they manufacturer their own heat pumps in-house or use private-label heat pumps in their GHP systems, they all provide contractor support and product warranty and service. So what are the differences?

1. Residential Warranties: There is quite a variation in warranties: from 5 to 12 years, even 20 years in one case. Labor is guaranteed by some companies, not by others. Most companies list warranties on their web sites or are pleased to provide information.

2. Training and installer accreditation: GHP manufacturers (with a few exceptions who offer turn-key installations) do not sell to the public. They rely on local contractors and installers who must be trained and accredited. When a manufacturer provides accreditation courses, it adds more qualified installers into the field who might well be more apt to use that manufacturer's product. It is a smart strategy. Of course, there are other sources for accreditation.

3. Efficiency Certification to Standards: Most GHP manufacturers become Energy Star Partners. In order to use the Energy Star label on heat pumps, a Partnership requires a commitment to use third-party verification testing through an EPA-certified organization, which in turn, standardizes the certification. They also must supply EPA with shipping data on the number of heat pumps shipped by category. All US companies listed next are Energy Star Partners.

(DX) *Companies offering Direct Exchange (DX) systems*

US Geothermal Residential Heat Pump Manufacturers

 Advanced Geothermal Technology
PO Box 6469
Reading, PA 19610
www.advgeo.com
(610) 736-0570
Direct Exchange (DX) systems, on-demand hot water.
Warranty: 5 years on compressor, 1 year on other parts, no labor.

Bard Manufacturing Co.
1914 Randolf Drive
Bryan, OH 43506
www.bardhvac.com
(419) 636-1194
Climate control, air conditioners, oil furnaces and heat pumps
Warranty: 5 years on parts including compressor.

Bosch Thermotechnology Corp.
50 Wentworth Ave.
Londonderry, NH 03053
(603) 552-1100
www. bosch-climate.us
Warranty 1 year parts and labor

Bryant Heating & Cooling Systems
7310 W. Morris St.
Indianapolis, IN 46231
www.bryant.com
(800) 428-4326
Warranty: 10 years on compressor and major parts; no labor; 5 years on other parts; extended warranty available.

Carrier Corporation
One Carrier Place
Farmington, CT 06032
www.residential.carrier.com
(800) CARRIER
Add-on water heating component.
Warranty: 10 years on compressor; 5 years on other parts.

Climate Master
7300 SW 44th St.
Oklahoma, OK 73179
www.climatemaster.com
(405) 745-6000
Warranty: 10 years on compressor and refrigerant circuit parts; 5 years on other parts; 5-year service labor allowance compressor and refrigerant circuit parts; optional extended warranty. Savings calculator.

Eagle Mountain
4376 Bristol Valley Rd
Canadaigua, NY 14424
www.eagle-mt.com
(800) 572-7831
Integrates wind, solar, geothermal. Turnkey installation kits.
Warranty: 5 years on parts and labor for refrigerant circuit; 2 years on other parts and labor.

 EarthLinked Technologies, Inc
4151 S. Pipkin Rd
Lakeland, FL 33811
www.earthlinked.com
(863) 701-0096
Direct Exchange (DX) systems, full demand hot water.
Warranty: 5 years on parts and labor for compressor parts and optional cathodic installation, 2 years on other parts and labor, 20 years on parts and 5 years on labor of earth loops.

ETA
(Earth to Air Systems)
123 SE Parkway Court
Franklin, TN 37064
www.earthtoair.com
(615) 595-2888
> Direct Exchange (DX) systems.
> *Warranty*: 5 years, plus extended
> warranty available.

FHP Manufacturing
601 NW 65th Court
Ft. Lauderdale, FL 33309
www.fhp-mfg.com/residential
(954) 776-5471
> Company purchased by Bosch of
> Germany in 2007. Savings calculator.
> *Warranty*: 5 years on compressor,
> 1 year other parts.

 Free Source Energies LLC
PO Box 122
Redwood Falls, MN 56283
www.freesourceenergies.com
(507) 644-2340
> Warranty: 10 years

GeoComfort
11699 N. Monique Rd
Northport, MI 49670
www.geocomfort.com
(888) 436-3783
> Website savings calculator,
> GeoSmart financing offered,
> GeoAnalyst software.
> *Warranty:* 10 years on compressor,
> 5 years on other parts, extended
> warranty available.

 GeoEnergy of New York
1520 Ocean Ave
Bohemia, NY 11716
www.geonrgny.com
(631) 319-0390
> Uses unique hybrid GeoColumn for
> DX and water systems.
> *Warranty:* 10 years, geothermal
> product only.

GeoMaster, LLC
3512 Cavalier Court
Ft. Wayne, IN 46808
www.geoexcel.com
(260) 484-4433
> Energy analysis and sizing software.
> *Warranty:* 10 years on refrigerant,
> circuit parts; 5 years on blower
> parts; 1 year on other parts/labor.

GeoSystems, LLC
Brands: Econar, HydroHeat
7550 Meridian Circle
Maple Grove, MN 55369
www.geosystemsghp.com
(800) 432-6627
> *Warranty:* 5 years on parts/labor;
> extended 10-year warranty available.

Heat Controller, Inc.
1900 Wellworth Ave
Jackson, MI 49203
www.heatcontroller.com
(517) 787-2100
> *Warranty:* 12 years for compressor,
> 6 years on other parts, no labor.

Hydron Module
(Enertech Manufacturing)
41659 256th St
Mitchel, SD 57301
www.hydronmodule.com
(800) 720-1724
> Savings calculator, GeoAnalyst
> software.
> *Warranty:* 10 years on sealed sys-
> tems, 5 years on electrical & labor.

Hydro Temp Corporation
PO Box 566
3636 Hwy 67 S
Pocahontas, AR 72455
www.hydro-temp.com
(800) 382-3113
    Priority on-demand hot water system available; kWh meter standard.
    Warranty: 5 years on compressor, 1 year on other parts.

Klimaire, Inc
7909 NW 54th St.
Miami, FL 32166
www.klimaire.com
(305) 593-8358
    Offering advanced infinitely variable compressor.
    Warranty: 5 years on compressor, 1 year on other parts, no labor.

Q Energy Systems, Dane Manufacturing
214 E. Maine St
Dane, WI 53525
www.qenergysystems.com
(608) 849-5921
    Wi-Fi real-time fault monitoring.
    Warranty: 10 years, geothermal product only.

Rheem-Rudd Manufacturing
101 Bell Rd
Montgomery, AL 36117
www.ruud.com
(334) 260-1500
    Warranty: 10 years on parts, HVAC products.

Sub Terra Energy Systems (DX)
24315 NE Dayton Ave
Newberg, OR 97132
www.subterraenergysystems.com
(503) 487-6326
    Direct Exchange (DX) systems, including installation only in Oregon and SW Washington.
    Warranty: 10 years on parts & labor.

TETCO Geothermal
2506 S. Elm St
Greenville, IL 62246
www.tetcogeo.com
(618) 669-9011
    Savings calculator, GeoSmart financing offered,
    Warranty: 5 years on major parts, 2 years on other parts and labor.

Trane
2 Corporate Wood Dr
Bridgeton, MO 63044
www.trane.com
(800) 945-5884
    Warranty: 10 years, HVAC products.

WaterFurnace International
9000 Conservation Way
Fort Wayne, IN 46809
www.waterfurnace.com
(800) 436-7283
    Savings calculator, Geolink sizing software.
    Warranty: 10 years on parts plus labor allowance, 5 years on unit accessories.

Canadian Residential Heat Pump Manufacturers

Boreal Geothermal, Inc.
785 Amherst St
Montreal, Quebec J2L 2J5
www.boreal-geothermal.ca
(450) 534-0203
Warranty: 5 years replace or repair all internal components.

Enertran Technology, Inc. (ETI)
5 Commerce Road
Orangeville, Ontario L9W 3X5
www.enertran.ca
(800) 941-0053
On demand hot water, plus systems to eliminate odor and humidity in very air tight homes.
Warranty: 5 years on compressor, installer must warrant installation for 10 months.

Geofinity Manufacturing
19050 25th Ave
Surrey, BC V3S 3V2
www.geofinitymanufacturing.com
(604) 536-4544
Warranty: 10 years

Geoflex Systems, Inc.
1069 Clarke Rd
London, Ontario N5V 3B3
www. geoflexsytems.com
(519) 488-1653
Warranty: 5 years on compressor, 2 years other parts; labor allowance.

GeoSmart Energy
290 Pinebush Road
Cambridge, Ontario N1T 1Z6
www. geosmartenergy.com
(519) 624-0400
GeoSmart Energy Academy provides training.
Warranty: 10 years on parts.

Maritime Geothermal Ltd
PO Box 2555
Petitcodiac, NB E4Z 6H4
www. nordicghp.com
(506) 756-8135
Nordic line of heat pumps, includes Direct Exchange (DX) systems.
Warranty: 5 years on all internal parts, labor allowance, extended warranty available.

Northern Heat Pump
3-201 South Railway Ave
Winkler, Manitoba R6W 1JB
www.northernheatpump.com
(204) 325-9772
ROI calculator on website.
Warranty: 5 years, 10 years optional.

PolarBear Geothermal Systems, Inc.
18 Crown Steel Dr
Markham, Ontario L3R 9X8
www. polarbearwshp.com
(888) 485-5854
Warranty: 5 years on parts/labor.

State Energy Offices

Alabama Energy Division
(334) 242-5290
*www.adeca.alabama.gov/
Energy*

Alaska Energy Authority
(907) 771-3000
Toll Free in AK: (888) 300-8534
www.aidea.org/aea

Arizona Energy Office
(602) 771-1137
www.azcommerce.com/Energy

Arkansas Energy Office
(800) 558-2633
www.arkansasenergy.org

**California Energy
Commission Renewable
Energy Programs**
(916) 654-4058
www.energy.ca.gov/renewables

Colorado Energy Office
Recharge Colorado
(303) 866-2100
www.rechargecolorado.com

Connecticut Energy Office
(860) 418-6200
www.ct.gov/opm

Delaware Energy Office
(302) 735-3480
www.dnrec.delaware.gov/energy

**District of Columbia Energy
Office**
(202) 673-6700
www.ddoe.dc.gov

**Florida Energy & Climate
Commission**
(850) 487-3800
www.dep.state.fl.us/energy

**Georgia State Energy
Program**
(404) 584-1000
*www.gefa.georgia.gov/
state-energy-program*

Hawaii Energy Office
(808) 587-3807
*www.hawaii.gov/dbedt/info/
energy*

**Idaho Office of Energy
Resources**
(208) 332-1660
www.energy.idaho.gov

**Illinois Bureau of Energy &
Recycling**
(217) 785-3416
*www.ilbiz.biz/dceo/Bureaus/
Energy_Recycling*

**Indiana Office of Energy
Development**
(317) 232-8939
www.in.gov/oed

**Iowa Office of Energy
Independence**
(515) 725-0431
www.energy.iowa.gov

Kansas Energy Office
(785) 271-3100
www.kcc.state.ks.us/energy

Kentucky Dept. for Energy Development and Independence
(502) 564-7192
www.energy.ky.gov

Louisiana Energy Office
(225) 342-1399
www.dnr.louisiana.gov
(Look under Energy)

Efficiency Maine
(866) 376-2463
www.efficiencymaine.com

Maryland Energy Office
(410) 260-7655
www.energy.maryland.gov

Massachusetts Department of Energy Resources
(617) 626-7300
www.magnet.state.ma.us/doer

Michigan Energy Office
(517) 241-6228
www.michigan.gov/dleg
(Look under Inside DELEG / Energy Systems)

Minnesota Office of Energy Security
(651) 296-5175
www.energy.mn.gov

Mississippi Energy Division
(601) 359-6600
www.mississippi.org
(Look under Energy)

Missouri Division of Energy
(573) 751-3443
www.dnr.mo.gov/energy

Montana Energy Office
Energize Montana
(406) 841-5200
www.deq.mt.gov/energy

Nebraska Energy Office
(402) 471-2867
www.neo.ne.gov

Nevada State Office of Energy
(775) 687-1850
www.energy.state.nv.us

New Hampshire Office of Energy and Planning
(603) 271-2155
www.nh.gov/oep

New Jersey's Clean Energy Program
(609) 777-3300
www.njcleanenergy.com

New Mexico Energy Conservation and Management Division
(505) 476-3310
www.emnrd.state.nm.us/ecmd

New York State Energy Research and Development Authority
(518) 862-1090
www.nyserda.org

North Carolina State Energy Office
(919) 733-2230
www.energync.net

North Dakota Office of Renewable Energy & Energy Efficiency
(701) 328-5300
www.communityservices. nd.gov/energy

Ohio Energy Resources Division
(614) 466-6797
www.development.ohio.gov/ Energy

Oklahoma State Energy Office
(405) 815-6552
www.okcommerce.gov/
State-Energy-Office/
SEO-Overview

Oregon Dept. of Energy
(503) 378-4040
www.oregon.gov/energy

Pennsylvania Office of Energy & Technology Deployment
(717) 783-0540
www.depweb.state.pa.us/energy

Rhode Island Office of Energy Resources
(401) 574-9100
www.energy.ri.gov

South Carolina Energy Office
(803) 737-8030
www.energy.sc.gov

South Dakota Energy Management Office
(605) 773-3899
*www.state.sd.us/boa/ose/OSE_*
Statewide_Energy.htm

Tennessee Office of Energy Policy
(615) 741-2994
*www.tn.gov/ecd/CD_office_*
energy_policy.html

Texas State Energy Conservation Office
(512) 463-1931
www.seco.cpa.state.tx.us

Utah State Energy Program
(801) 537-3300
www.geology.utah.gov/sep

Vermont State Energy Office
(802) 828-2811
www.publicservice.vermont.gov
(Renewables & Efficiency)

Virginia Division of Energy
(804) 692-3200
www.dmme.virginia.gov/
divisionenergy

Washington State Energy Policy Division
(360) 725-3118
www.cted.wa.gov
(Look under Energy Policy)

West Virginia Division of Energy
(304) 558-2234
www.wvcommerce.org/energy

Wisconsin Office of Energy Independence
(608) 261-6609
www.energyindependence.wi.gov

Wyoming State Energy Program
(307) 777-2800
www.wyomingbusiness.org/
business/energy.aspx

Glossary

Btu. British thermal unit; the amount of energy needed to raise the temperature of one pound of water by one degree Fahrenheit, at one atmosphere. **Btu/h** is the number of Btus produced in one hour.

closed-loop system. The same water or refrigerant/water solution is constantly recirculated within sealed pipes or tubes.

direct exchange (DX) system. Sometimes referred to as DGX, Direct Ground eXchange is a closed-loop GHP system which uses two loops instead of the standard three-loop approach.

compressor. The main component of a heat pump system; it increases the pressure and temperature of the refrigerant while reducing its volume, causing the refrigerant to move through the system.

condenser. A heat exchanger in which hot, pressurized (gaseous) refrigerant is reduced to a hot pressurized liquid by transferring heat to a cooler substance (air, water, earth).

COP (coefficient of performance). A measure of GHP efficiency in the heating mode; the total Output heating capacity (in Btu) divided by the electrical power Input. Numbers range from 3 to 5 (300% to 500%). The higher the COP, the more efficient the system.

desuperheater water heater. A device for recovering "excess heat" from the compressor discharge gas of a heat pump or central air conditioner; for use in heating or preheating domestic water.

direct-use geothermal. Geothermal heating systems that use naturally heated underground water, such as hot springs or reservoirs.

EER (energy efficiency ratio). The measure of GHP efficiency in the cooling mode; the total Output cooling capacity (in Btu/hour) divided by the power Input (in watts). Numbers range from 9 to 31.

efficiency. The ratio of the useful work performed by a machine or in a process to the total energy expended or heat taken in.

energy. The ability to do work; comes in many forms, including electrical, mechanical, heat, nuclear, and light.

fossil fuels. Carbon- and hydrogen-laden fuels (now combustible) formed underground from the remains of long-dead plants and animals, such as crude oil, natural gas, propane, and coal.

geoexchange system. An electrically powered heating and cooling system for interior spaces, using the earth as a heat source and heat sink. *See* GHP.

geothermal energy. Heat generated by natural processes within the earth. The chief energy resources are hot dry rock, magma (molten rock), hydrothermal (water/steam from geysers and fissures), and geopressure (water saturated with methane under tremendous pressure at great depths).

GEP. Geothermal Electric Power converts heat energy from deep in the earth into electrical energy for large-scale utility operations.

GHP (geothermal heat pump).
A system that uses the earth as a
heat source and heat sink; used for
residential and commercial heating
and cooling.

green building. A building or construction method that makes use
of sustainable, renewable materials and principles and strives for
energy efficiency.

ground-source heat pump. *See*
GHP, geothermal heat pump.

heat. The total thermal energy in
a substance; measured in calories,
Joules or Btu. It can be transferred
from one substance to another.

heat exchanger. A device designed
to transfer heat between two physically separated fluids or mediums of
different temperatures.

heat pump. A mechanical device
that transfers heat from a colder
area to a hotter area. They are clas-
sified as either air-source or water-
source units.

horizontal closed-loop system.
Plastic piping or tubing that is
buried horizontally, about 6 to 8
feet below the surface. The closed
loop recirculates the antifreeze/
water solution for the ground-loop
heat transfer.

HVAC and HVAC/R. Heating,
ventilation, and air conditioning,
(and refrigeration); acronyms used
to describe such systems and the
people who work with them.

Ideal Gas Law. In any closed, fixed-
volume system, the Pressure (P) is
proportional to the Temperature
(T). If P increases (or decreases),
then T must increase (or decrease).

kilowatt (kW). Unit of electrical
power equal to 1,000 watts.

kWh (kilowatt-hour). A standard
unit of electricity production and
consumption. One kWh = 3,412 Btu.

latent heat transfer. Temperature
remains unchanged during this heat
transfer, but energy is still being
absorbed and stored. The substance
changes physical state or phase (i.e.
liquid to gas, or gas to liquid).

Laws of Thermodynamics.
Describe how fundamental physical
quantities (temperature, energy,
and entropy) behave under various
circumstances in thermodynamic
systems.

net zero. the total amount of energy
used by the building on an annual
basis is roughly equal to the amount
of renewable energy harvested
onsite.

open-loop systems. The ground-
water is not recirculated; it is
returned back to the source.

payback. the time it takes to recover the difference in cost between
two different HVAC systems, taking
into account the energy and maintenance cost savings.

phase change. A change from
one state (solid or liquid or gas) to
another state without a change in
chemical composition.

photovoltaic (PV). The technology
of converting sunlight directly into
electricity through the use of photovoltaic (solar) cells. Photovoltaic
cells are made primarily of silicon,
which responds electrically to the
presence of sunlight, a phenomenon
known as the photovoltaic effect.

pressure. The force exerted on a given area.

refrigerant. A material used in a refrigeration cycle which undergoes phase changes from a gas to a liquid and back again.

refrigeration cycle. A thermodynamic process whereby a refrigerant accepts and rejects heat in a repetitive sequence.

renewable energy. Any form of energy that derives from sustainable resources that can be naturally renewed, such as wind, solar, geothermal.

reversing valve. A component in a GHP system that changes the direction of refrigerant flow; heating can be switched to cooling, or vice versa.

sensible heat transfer. Affects the temperature of a substance while heat energy is being absorbed and temporarily stored.

Slinky coils. Flattened, overlapped plastic pipe in circular coiled loops concentrate the heat transfer area into a smaller volume, reducing land requirements for the ground loop.

solar electricity. Electricity generated directly by sunlight via photovoltaic cells.

supplemental heating. A separate heating system used when extra heat is needed to moderate indoor temperatures.

sustainability. Relating to, or being a method of harvesting or using a resource so that the resource is not depleted or permanently damaged.

tankless water heating. A backup domestic water heating system often used with GHPs; heats water on-demand (as it is needed).

temperature. The measure of the average molecular motion of a substance; measured in degrees Fahrenheit (F) or Celsius (C).

TXV, thermostatic expansion valve. Performs the opposite function of a compressor by allowing the high pressure refrigerant to expand, thus lowering the pressure and temperature.

vertical closed loops. Small diameter holes are drilled into the earth to hold the plastic pipping that recirculates the antifreeze/water solution for the ground loop heat transfer.

water-to-air system. Transfers heat from the GHP ground loop (water) to the home's air distribution system (air) for heating and cooling.

water-to-water system. Transfers heat from the ground loop (water) to the home's hot water heating system.

watt (W). Unit of electrical power used to indicate the rate of energy produced or consumed by an electrical device.

Zero Energy Home (ZEH). A general term applied to a home that has net zero energy consumption and zero carbon emissions annually. A ZEH combines high levels of energy efficiency with state-of-the-art renewable energy systems to annually return as much energy to the utility as it takes from the utility.

Index

"i" represents illustrations or photos

Acknowledgments

There is a popular belief that a writer as an author creates a book. That is not true. An author creates a manuscript. That manuscript is the heart and soul of a book, but it is not a book.

It requires an experienced editor, especially in a non-fiction genre, to pull together the words, photos, graphs, and sidebars and then paginate them in a cohesive way that is pleasing to the eye, has a logical flow, meets literary standards and is actually readable. An index must be created, possibly a glossary, inconsistencies eliminated, and a check for possible copyright infringement. When printed it becomes a book. Of course, at this point the author always calls it his or her book.

A geothermal heat pump is a sort of metaphor for a manuscript. The heat pump manufacturer pulls together past knowledge and standards, then designs and creates a piece of hardware—a heat pump. That is not a residential heating system although it is the heart of that system. It takes an experienced contractor to put together the various pieces—the ground loop, the wiring, the plumbing, the controls, and a room heat delivery subsystem in a cohesive way that has a logical flow, and is actually workable. Only then, when the thermostat is turned on and heat enters the room does it become a residential heating system.

But LaVonne Ewing, as my publisher and editor, carried this process one step further. She was familiar with the subject and she contributed interesting ideas and offered suggestions during the manuscript creation process. My end was to convert all of these ideas into meaningful words and integrate them into the story I am telling. This was live-wire stuff, energizing. The collaboration resulted in a far better book, in my opinion.

I have not spent my working life in the geothermal heat pump business; I do, however, have some experience as a homeowner in installing and living with a GHP. And I know something about the physics of heat energy transfer. But I had to go way beyond that to adequately tell the GHP story. In addition to the power of the internet, I found a network of people actively working in the field who were exceedingly helpful and gracious in answering questions, contributing relevant material, and commenting on the manuscript. I want to especially acknowledge a few:

I would like to thank Gerald McClain of IGSHPA, an Oklahoma State University Professor Emeritus, for his critique and valuable suggestions for the book; John Geyer, Certified Geothermal Designer, trainer and consultant in Vancouver for his contributions; and Ed Lohrenz in Canada, an IGSHPA Certified Instructor of installer and design workshops and a partner in the consulting firm GEO-Xergy Systems, Inc.

Chuck Russo, my GHP contractor, is one of those very qualified and experienced installers that the world needs more of. He has installed hundreds of GHP systems in the New York, Massachusetts and Vermont area. He was patient with me as I learned how to deal with this new system. He knew that I had never even seen a GHP until it was carted into my basement. So, he was not too impressed when he learned that I was working on a GHP book. That is until he read my Table of Contents and learned that I had a publisher. That did it—he asked for a copy of my manuscript, reviewed it and made helpful suggestions. Thank you, Chuck, for being among the first installers to see the value in giving copies of my book to prospective customers.

Paul Auerbach and Diana Gaeta of Total Green Geothermal (*www.totalgreenus.com*) in Monroe, New York, have provided data on case study examples from their files and important information on DX geothermal systems in which they specialize. Paul has a DX system in his own home and as a contractor and installer, his company installed over 150 DX systems as of 2014. In addition to GHPs, they are also experts at the larger renewable energy picture, including a building's thermal envelope and solar/wind power.

I would like to also thank those homeowners who permitted me to include their homes here as interesting case studies. Then, of course, my thanks to those companies (in addition to Total Green Geothermal above) who provided other case studies around the continent: Robert Hoffman of Hoffman Brothers Heating and Air Conditioning in St. Louis, MO; Randy Waylett of Northern Heat Pump in Manitoba; and Don Murray of Enertran Heat Pumps in Ontario.

A special thanks to Will Suckow for his drawings that bring a certain charm and an important cohesiveness to the story.

Most importantly, I want to thank Lawrence Muhammad, a solid professional, for participating in this project. He brings a depth of experience and knowledge that is extremely valuable.

Author Don Lloyd

Don Lloyd is an electrical engineer now retired from Honeywell after 35 years in marketing, engineering and project management. He has extensive experience in taking complex subjects (such as laser gyroscopes or mosaic infrared detectors) and explaining how they work in terms understood by the general public. He is at his best when giving lectures describing how geothermal heat pumps work and why they are the ideal heating and cooling solution for homeowners. Prior to entering the geothermal world, he operated his own software consulting business as a Microsoft Certified Software Professional. He also served as an officer in the US Army.

Don lives in the Hudson River Valley region of New York State with his wife, artist Martha Lloyd, in their new home where the oil trucks pass by but never stop. "Installing a geothermal heat pump was one of the best decisions I ever made!"

Technical Advisor Lawrence A. Muhammad

Lawrence A. Muhammad is an International Ground Source Heat Pump Association (IGSHPA) Certified National Trainer, having conducted numerous trainings throughout the US. As a community college professor, he teaches a wide variety of geothermal courses at three colleges in Michigan.

Over the years, Lawrence has been invited to speak at several renewable energy and sustainability conferences. Recently he traveled to Jamaica to discuss geothermal applications which can benefit the Jamaican/CARICOM economy. Establishing "local economies" using hybrid renewable energy baseload sources is as important to Lawrence as it is to workforce development entities and urban communities.

Putting his teaching and training skills to work, Lawrence is also a licensed residential builder, an IGSHPA Accredited Installer, a NATE Proctor, and the designer/developer of the apprenticeship program for the Geothermal Innovations Gii4 proprietary heat exchanger. His geothermal training company, Geo NetZero, is committed to providing a skilled and qualified workforce throughout America.

www.GeoPowerful.com

100% solar & wind powered since 1999
PIXYJACK PRESS INC

*Visit **PixyJackPress.com** to view all of our renewable energy titles and to place educational / wholesale orders.*